AF347528

Advances and Challenges in Organic Electronics

Advances and Challenges in Organic Electronics

Editors

Frédéric Dumur
Fabrice Goubard

MDPI • Basel • Beijing • Wuhan • Barcelona • Belgrade • Manchester • Tokyo • Cluj • Tianjin

Editors
Frédéric Dumur
Aix-Marseille University
France

Fabrice Goubard
Cergy-Pontoise University
France

Editorial Office
MDPI
St. Alban-Anlage 66
4052 Basel, Switzerland

This is a reprint of articles from the Special Issue published online in the open access journal *Materials* (ISSN 1996-1944) (available at: https://www.mdpi.com/journal/materials/special_issues/ optoelectronic_materials_devices).

For citation purposes, cite each article independently as indicated on the article page online and as indicated below:

LastName, A.A.; LastName, B.B.; LastName, C.C. Article Title. *Journal Name* **Year**, *Volume Number*, Page Range.

ISBN 978-3-0365-0282-3 (Hbk)
ISBN 978-3-0365-0283-0 (PDF)

Contents

About the Editors

Frédéric Dumur received his Ph.D. in chemistry in 2002 from the University of Angers under the supervision of Professor Pietrick Hudhomme. Following completion of his Ph.D., he was successively a Post-Doctoral Research fellow at the University of Groningen (the Netherlands), Reims Champagne-Ardennes (France) and Versailles Saint-Quentin en Yvelines (France). In 2008, he joined the Faculty of Sciences at the Aix-Marseille University, where he is currently an Associate Professor. His research interests include the synthesis of phosphorescent dopants for OLEDs and photoinitiators of polymerization.

Fabrice Goubard is a full-time professor of Chemistry in the LPPI at the CY Cergy Paris Université. His expertise extends to both hybrid and organic photovoltaic applications. He received his Ph.D. (1995) degree in Solid Chemistry from P. et M. Curie University (Paris, France). He joined the LPPI in 1996 as an assistant professor. His research currently deals with pi-conjugated molecular glasses for hybrid photovoltaic devices (notably perovskite solar cells), and stabilization of organic photovoltaic cells with the IPN network.

Preface to "Advances and Challenges in Organic Electronics "

Organic Electronics is a rapidly evolving multidisciplinary research field at the interface between Organic Chemistry and Physics. Organic Electronics is based on the use of the unique optical and electrical properties of π-conjugated materials that range from small molecules to polymers. The wide activity of researchers in Organic Electronics is testament to the fact that its potential is huge and its list of potential applications almost endless. Application of these electronic and optoelectronic devices range from Organic Field Effect Transistors (OFETs) to Organic Light Emitting Diodes (OLEDs) and Organic Solar Cells (OSCs), sensors, etc. We invited a series of colleagues to contribute to this Special Issue with respect to the aforementioned concepts and keywords. The goal for this Special Issue was to describe the recent developments of this rapidly advancing interdisciplinary research field. We thank all authors for their contributions.

Frédéric Dumur, Fabrice Goubard
Editors

Review

Recent Advances on Nitrofluorene Derivatives: Versatile Electron Acceptors to Create Dyes Absorbing from the Visible to the Near and Far Infrared Region

Guillaume Noirbent * and Frédéric Dumur *

Aix Marseille Univ, CNRS, ICR UMR 7273, F-13397 Marseille, France
* Correspondence: guillaume.noirbent@outlook.fr (G.N.); frederic.dumur@univ-amu.fr (F.D.); Tel.: +33-(0)4-9128-2748 (F.D.)

Received: 11 November 2018; Accepted: 29 November 2018; Published: 30 November 2018

Abstract: Push–pull dyes absorbing in the visible range have been extensively studied so that a variety of structures have already been synthesized and reported in the literature. Conversely, dyes absorbing in the near and far infrared region are more scarce and this particularity relies on the following points: difficulty of purification, presence of side-reaction during synthesis, low availability of starting materials, and low reaction yields. Over the years, several strategies such as the elongation of the π-conjugated spacer or the improvement of the electron-donating and accepting ability of both donors and acceptors connected via a conjugated or an aliphatic spacer have been examined to red-shift the absorption spectra of well-established visible dyes. However, this strategy is not sufficient, and the shift often remains limited. A promising alternative consists in identifying a molecule further used as an electron-accepting group and already presenting an absorption band in the near infrared region and to capitalize on its absorption to design near and far infrared absorbing dyes. This is the case with poly(nitro)fluorenes that already exhibit such a contribution in the near infrared region. In this review, an overview of the different dyes elaborated with poly(nitro)fluorenes is presented. The different applications where these different dyes have been used are also detailed.

Keywords: fluorene; nitrofluorene; Knoevenagel reaction; near infrared absorption; push–pull chromophore; poly(nitro)fluorene

---

1. Introduction

The reciprocal influence of an electron-donating and an electron-accepting group is capable to give rise to structures with fascinating properties. Optically, the first manifestation of this mutual interaction is a color change when the two partners are mixed, resulting in the presence of an additional absorption band which is not detectable in the absorption spectra of the initial partners considered separately. This absorption band corresponds to the charge transfer (CT) interaction where part of the electronic density is transferred from the electron-rich to the electron-deficient partner. Depending on the fact that the two moieties are (or not) connected to each other, this CT transition can be intermolecular or intramolecular. Historically, intermolecular charge transfer complexes have been extensively studied, focused on the metallic conductivity resulting from the grinding of two insulating organic compounds, i.e., tetrathiafulvalene (TTF) and tetracyano-quinodimethane (TCNQ) [1]. Following this pioneering work, the possibility to develop superconductors at low temperatures has driven extensive research efforts, and the emergence of the Bechgaard salts [2]. If these structures are attractive for their electrical properties, the resulting charge transfer complex renders these structures insoluble, drastically limiting the scope of applicability. Conversely, materials exhibiting an intramolecular charge transfer (ICT) i.e., meaning that the electron donor is connected to the electron acceptor by mean of a spacer (aliphatic or conjugated) are not necessarily insoluble so that these structures found widespread applications ranging from dye-sensitized solar cells (DSSCs) [3], nonlinear optical applications [4,5], solvatochromic probes [6–8], photochromes [9], single component semiconductors [10–12], electrochromes [13,14], and piezochromes [15]. Typically, these compounds are composed of a π-conjugated system with an electron-donating, and an electron-accepting part connected at both sides of the conjugated spacer [16–19]. If the number of groups used as electron donors cannot be calculated anymore in regards to the diversity of structures, the number of electron acceptors that can be covalently linked to electron donors is more limited, as exemplified by the list of molecules 1–17 presented in the Figure 1. Notably, malononitrile **1** [20], indanedione derivatives **2** [21], (thio)barbituric derivatives **3** [22], Meldrum derivatives **4** [23], pyridinium **5** [24], methyl-containing tricyanofurans **6** [25], substituted tricyanopropenes **7** [26], pyran derivatives **8** and **9** [27,28], 1,1,3-tricyano-2-substituted propenes **10** [29], isoxazolones **11** [30], hydantions and rhodanines **12** [31], pyrazines **13** [32], dicyanoimidazoles **14** [33], benzo[*d*]thiazoliums **15** [34], benzo[*d*]imidazoliums **16** [35], and dicyanovinyl-thiophen-5-ylidenes **17** [36] can be cited as the most common acceptors.

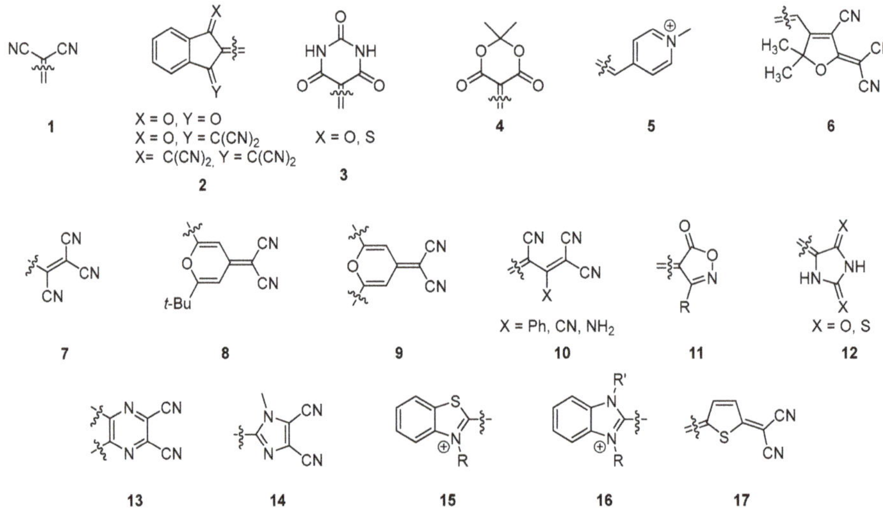

Figure 1. Chemical structures of common electron acceptors.

For all these acceptors and irrespective of the electron donors, the intramolecular charge transfer band is centered in the visible range and this latter can be displaced towards the near infrared region if the length of the π-conjugated spacer introduced between the two partners is extended as much as possible. However, the extension of the π-conjugated spacer is often a hard work from a synthetic point of view and alternatives are actively researched [37]. Considering that for numerous applications, an absorption in the near-infrared region is required and face to the fact that all the above-mentioned electron acceptors are not sufficiently electro-deficient to inherently position the charge transfer band in the near infrared (NIR) region without taking recourse to the elongation of the π-conjugated spacer, electron acceptors already displaying an absorption band in the NIR region are actively researched. In this field, and to the best of our knowledge, only (poly)nitrofluorenes such as dinitrofluorene **19** [38], trinitrofluorene **20** and tetranitrofluorene (TNF) **21** exhibit such an absorption band (see Figure 2). It has to be noticed that the absorption band detected in the NIR for **19–21** is not detectable in **18**. Based on this unique property, numerous push–pull dyes displaying an absorption peak in the NIR region have been designed. If (polynitro)fluorenone derivatives have been extensively studied for the design of intermolecular charge transfer complexes [39], the presence of the carbonyl function in fluorenone totally impede this structure to be used for the design of push–pull molecules by connecting the electron withdrawing/releasing groups to the carbon inserted between the two aromatic rings. Fluorenone derivatives will not be discussed in this review (see Figure 3). Indeed, by the presence of the activated CH_2 group standing between the two aromatic rings, poly(nitro)fluorene derivatives constitute candidates of choice for the synthesis in one step of push–pull molecules by a Knoevenagel reaction, when opposed to an electron donor comprising an aldehyde function (see Figure 3).

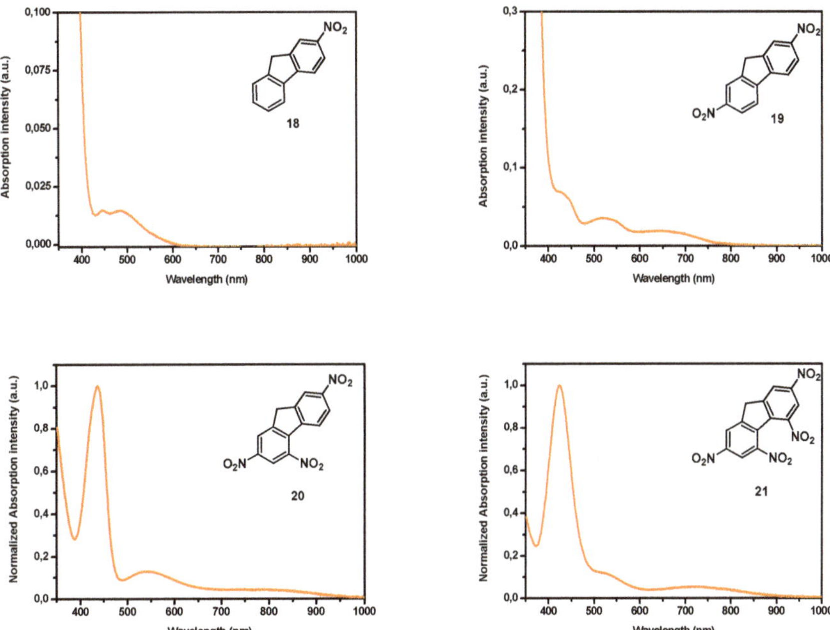

Figure 2. UV-visible absorption spectra of **18–21** in *N,N*-dimethylformamide (DMF).

As an interesting feature, when substituted with electron-withdrawing groups such as nitro groups, the methylene group standing between the two aromatic rings is sufficiently acid so that the Knoevenagel reaction on **20** and **21** can be carried out without any base, in non-toxic and polar solvents such as *N,N*-dimethylformamide (DMF). Principles of Green Chemistry can thus be applied to the synthesis of push–pull molecules (see Figure 3). Conversely, the methylene group in **18** and **19** is often not sufficiently activated to allow the Knoevenagel reaction to be carried out without a base and piperidine or pyridine are classically used.

Figure 3. Chemical structures of fluorene and fluorenone.

In this review, an overview of the different push–pull structures comprising (poly)nitrofluorenes as electron acceptors is presented. More precisely, four different aspects will be detailed. First, numerous push–push dyes have been obtained by functionalization of the 9-position of the fluorene acceptor. Second, a series of dyes has also been obtained by nucleophilic substitution of secondary amines on cyano-substituted fluorenes and this second approach constitutes the second most widely used synthetic procedure to access to poly(nitro)fluorene-based dyes. Third, the conceptual unimolecular rectifier proposed by Aviram and Ratner has been a source of inspiration for the design of numerous chromophores and a series of dyes has been designed based on this proposed structure. Finally, parallel to organic donors, organometallic donors have also been investigated, and it constitutes the last part of this review. To end, and parallel to the synthesis and the examination of the different photophysical properties of the dyes, the different applications justifying the design of these structures will be detailed.

2. Push–Pull Molecules Based on the Connection of Strong Electron Donors at the 9-Position of the Fluorene Acceptors

The first report mentioning the design of push–pull molecules with poly(nitrofluorenes) **23–26** was published in 1984 [40]. In this pioneering work, relatively weak electron donors were used since alkylthio groups were employed. By virtue of the electron-donating ability of the sulfur atom, an absorption extending until 480 nm could be obtained for **25** and **26**. Despite the weak electron-donating ability of alkylthiols, the strong electron-accepting ability of **19–21** could be evidenced, the standard conditions of thioacetalization of ketones being unable to provide the targeted molecules. Indeed, the charge transfer interactions existing between the alkylthiols and nitrofluoren-9-one derivatives hampered the classical condensation procedures to provide the push–pull structures **23–26** (See Figure 4).

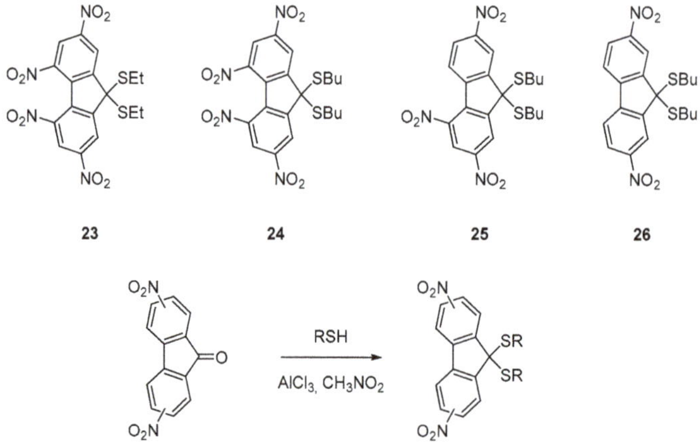

Figure 4. First push–pull molecules **23–26** based on poly(fluorene) acceptors.

The four molecules could only be obtained using an unusual procedure making use of aluminium chloride as a mediator for the reaction [41]. Using this strategy, **23–26** could be obtained with reaction yields ranging from 85 to 93%. Interestingly, a weak charge transfer interaction could be evidenced for **23–26**, the lowest-energy transition exhibiting an onset at 470 nm for **25** and **26** and 410 nm for **23** and **24** respectively. Weakness of the interaction between the sulfur donor and the nitro acceptors and thus the low electronic delocalization in the ground state was confirmed by cyclic voltammetry, all molecules exhibiting redox potentials comparable to that obtained for alkyl 2,4,5,7-tetranitrofluorene-9,9-dipropionates that does not possess any electron-donating groups. These molecules possessing good electron acceptors connected to electron donors can efficiently sensitize the photoconductivity of semiconducting polymers by forming charge transfer complexes with these latter, especially with carbazole-containing polymers [42–44], and this ability was examined with poly(*N*-vinylcarbazole) (PVK) which is a standard semiconducting polymer. This interaction was demonstrated by using two different techniques i.e., the space-charge-limited xerographic discharge and by an electrochemical time-of-flight (ETOF) method. It has to be noticed that the first commercially available organic photoconductor was based on a charge-transfer complex between tetranitrofluorene (TNF) and PVK, justifying the different studies devoted to this topic. As interesting feature, the spectral zone of photoconductivity of the carbazole-containing polymers can be selected by the appropriate choice of the push–pull molecule. By comparing the electron mobilities of **23–26** with that of the reference 2,4,7-trinitrofluoren-9-one, comparable transport properties could be determined [45–48], demonstrating the pertinence of the strategy for the design of photoconductive materials with poly(nitro)fluorene derivatives.

In contrast, by improving the electron donating ability of the donors, a significant red-shift of the absorption combined with the appearance of a second and new long-wavelength absorption band can be detected at lower energy than the intramolecular charge transfer. This specific band could be evidenced with the series of push–pull molecules **27–32** (see Figure 5) [49]. This additional absorption band is really a characteristic from the poly(nitro)fluorene acceptors. As electron donors, dithiolylidene and selenathiolylidene groups were used, these groups being among the best electron donors [50,51]. **27–32** were obtained by condensation of the corresponding dithiolylium **35**, **36** and selenathiolylium salts **37** with the electron acceptors **20**, **21**, **33** and **34** [52,53]. More precisely, due to the presence of multiple electron-accepting nitro groups on **20** and **21**, these latter are strong C-H acids [54] allowing the condensation in a highly polar solvent such as DMF to be carried out without any base. From a photophysical point of view, two charge transfer bands could be detected for the whole series **27–32**, the first one being detected between 446 and 500 nm and the second one between 587 and 633 nm respectively. Almost no influence of the heteroatom on the electronic transitions of **31** and **32** could be evidenced (see Table 1).

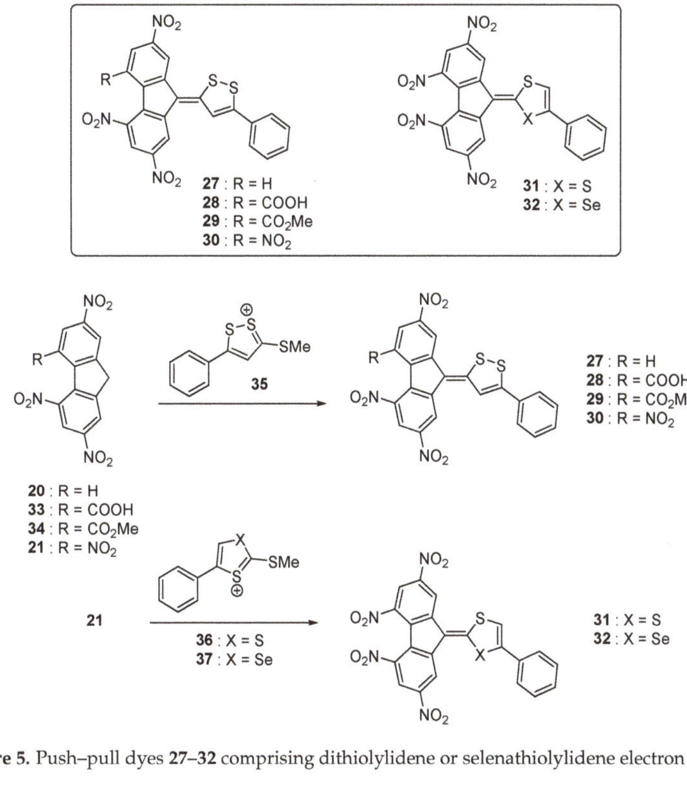

Figure 5. Push–pull dyes **27–32** comprising dithiolylidene or selenathiolylidene electron donors.

Table 1. UV-visible absorption characteristics of compounds **27–32** in *N,N*-dimethylformamide (DMF).

Compounds	27	28	29	30	31	32
ICT Bands (nm)	500, 587	492, 604	498, 613	509, 633	447, 599	446, 602

In sharp contrast, positions of the sulfur atoms clearly influence the ICT bands and a blue-shift of 60 nm for **31** relative to that of **30** was determined. Considering that the same electron acceptor is used for **30** and **31** (i.e., **21**), the enhanced electron donating ability of 1,2-dithiol-3-ylidene compared to that of 1,3-dithiol-2-ylidene was demonstrated. As attended, improvement of the electron accepting ability of the substituents in **27–30** resulted in a bathochromic shift of the ICT bands (see Table 1) and a bathochromic shift of the absorption maximum was also determined by increasing the solvent polarity. It thus evidences the higher polarity of the excited state for these compounds relative to that of their ground states. Intramolecular nature of the charge transfer was proved by protonation of the different molecules in sulfuric acid, resulting in the disappearance of the two ICT bands and the formation of colorless acidic solutions.

These different trends (lack of influence of the heteroatoms on the ICT maxima, positive solvatochromism) were confirmed in a subsequent study devoted to a series of 28 molecules (**27–32, 38–59**) (See Figure 6) [55] Several molecules of the former study (namely **27–32**) were revisited in the context of this new work with aim at establishing a structure-photophysical properties relationship.

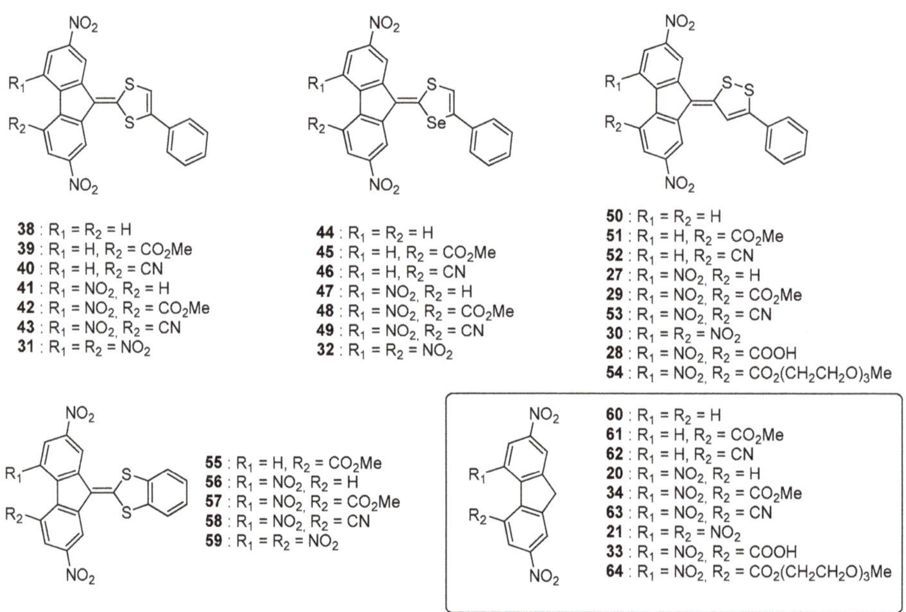

Figure 6. Push–pull dyes **27–32**, **38–59** using 1,2- and 1,3-dithiole, and 1,3-selenathiole as electron donors.

Nine acceptors **20, 21, 33, 34, 60–64** were examined and the synthesis of the different dyes were carried out using procedures similar to that previously reported, using **35–37** and **65** as the electron donors. New conclusions could be determined from this work examining a wide range of structures. First, the reaction yields significantly decreased by reducing the electron-accepting ability of R$_1$ and R$_2$ so that use of a base was required while using **60–62** as acceptors (see Figure 7). As specificity, **60–62** are not substituted with nitro groups, deactivating the methylene group of the fluorene acceptor. The intramolecular nature of the charge transfer complex was proved by dissolving the different compounds in sulfuric acid, resulting in the formation of colorless solutions by protonation of the electron-donating part. Reversibility of the process was also evidenced upon addition of water, regenerating the initial compounds.

Only acidic sensitive compounds comprising nitrile or ester groups could not be entirely recovered, these groups being partially hydrolysed in strongly acidic conditions. Intramolecular nature of the charge transfer band was also proved by UV-visible spectroscopy, with a perfect and linear concentration dependence of the absorbance of their absorption maxima.

As other finding, modification of the linkage of the benzene ring to the donor group drastically alter of the absorption properties, with a hypsochromic shift of the absorption peaks of about 70 nm when the aromatic ring is fused to the donor (see series of compounds **55–59** vs. the series of molecules **27–30, 50–54**). Examination of the solvatochromism proved to be quite complex, all the dyes exhibiting two intramolecular charge transfer bands originating from different transitions. (see Table 2).

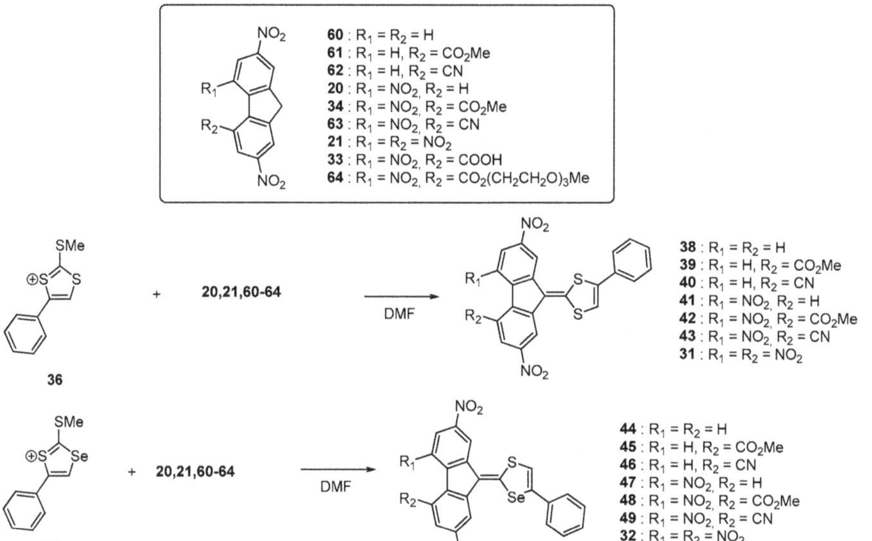

Figure 7. Synthetic routes to **27–32, 38–59**.

Considering that the electronic transitions involved in these different absorption bands are not the same, the solvatochromic behaviors of the two bands varied separately and opposite behaviors were sometimes found for the two bands of a same dye. While using correlations based on the basicity and/or the polarity of the solvent such as the Koppel–Palm correlation, no linear relationships could be obtained [56,57]. A similar behavior was observed while using more traditional multi-parameter polarity scales such as the normalized Reichardt's parameter (E^N_T) or the Dimroth–Reichardt $E_T(30)$ index [58]. Parallel to the solvatochromism, a thermochromism could be detected for all compounds, evidencing a modification of the solute-solvent interactions with the temperature [59,60]. Thus, a hypsochromic shift while increasing the temperature was determined for all compounds, with a more pronounced shift for the low energy ICT band. To illustrate this, a blue shift of 72 nm was observed in chlorobenzene for the second ICT band of **31** while increasing the temperature from 3 to 92°C whereas a blue shift of only 7 nm was detected for the high-energy transition.

Table 2. UV-visible absorption characteristics of compounds in dioxane.

Compounds	38	39	40	41	42	43	31
ICT Bands (nm)	406, 491	412, 492	418, 515	420, 525	426, 556	426, 552	434, 571
Compounds	**44** [1]	**45** [1]	**46**	**47**	**48**	**49**	**32**
ICT Bands (nm)	403, 490	410, 514	416, 516	422, 526	425, 556	425, 554	433, 571
Compounds	**50**	**51**	**52**	**27**	**29**	**53**	**30**
ICT Bands (nm)	460, 527	470, 557	470, 565	480, 572	490, 592	486, 593	492, 606
Compounds	**28** [2]	**55**	**56**	**57**	**58**	**59**	
ICT Bands (nm)	492, 604	498	513	542	549	425, 569	

[1] in acetone. [2] in DMF. ICT: intramolecular charge transfer.

To quantitatively examine the temperature effect, a linear dependence with inverse temperature could be evidenced. As attended, a negative halochromic behavior was detected, resulting from the protonation of the electron-donating dithiolylium. Interestingly, at certain concentration, an additional band in the 800–1000 nm region was detected for **31** and **32**, assigned to the formation of radical ion species. Precisely, the generation of radical species was assigned to bimolecular interactions between the neutral form regenerated upon dilution and the protonated form, resulting to an intermolecular electron transfer between the neutral and the cationic form, producing a radical and a radical cation following the Equation (1).

$$\mathbf{31 + 31\text{-}H^+ \leftrightarrow 31^+ + 31\text{-}H} \tag{1}$$

Electrochemical investigations revealed the different molecules to exhibit two redox processes, with two closely spaced reduction waves centered on the fluorene moieties, corresponding to the formation of the radical anion and dianion. For the series of compounds **32**, **44–49**, an additional reversible reduction process was detected at about -2.0 V (vs. Fc^0/Fc^+ couple) and this process was assigned to the formation of the radical trianion species. Investigation of the substitution effects on the fluorene acceptors revealed the difference between the two first reduction processes to be almost constant, whatever the substitution of the fluorene core was. Position of the ICT band is highly sensitive to the strength of the electron donors and acceptors that are selected for designing push–pull dyes but position of this transition can also be modified with lengthening the conjugated chain between the two partners. Influence of this parameter was notably examined with a series of chromophores bearing 1,3-dithiolylium donors [61]. The different dyes **66–97** were synthesized using a strategy comparable to that used for the former series (see Figure 8). **98–105** were used as the electron donors and **20**, **21**, **60–64** as the electron acceptors.

Upon elongation of the conjugated spacer, an extension of the highest occupied molecular orbital (HOMO) level over the donor part and the π-conjugated system is logically observed, resulting in a decrease of the HOMO-LUMO gap (where LUMO stands for lowest unoccupied molecular orbitals). Thus, comparison between **71**, **83**, and **90** revealed the second ICT band to shift from 544 to 588 and 636 nm respectively in 1,2-dichloroethane (see Table 3). Parallel to this, the molar extinction coefficient drastically increased from 9800 to 21,000 and 39,000 $dm^3 \cdot mol^{-1} \cdot cm^{-1}$ respectively upon elongation of the conjugated spacer. By electrochemistry, a cathodic shift could be detected (E_{red} (**71**) = -0.80 V, -1.02 V and E_{red} (**83**) = -0.83 V, -1.05 V vs. Ferrocene/Ferrocenium (Fc^0/Fc^+) couple), consistent with an improvement of the electron-donating ability in **83** and thus a lower ability for the fluorene moiety to accept electrons.

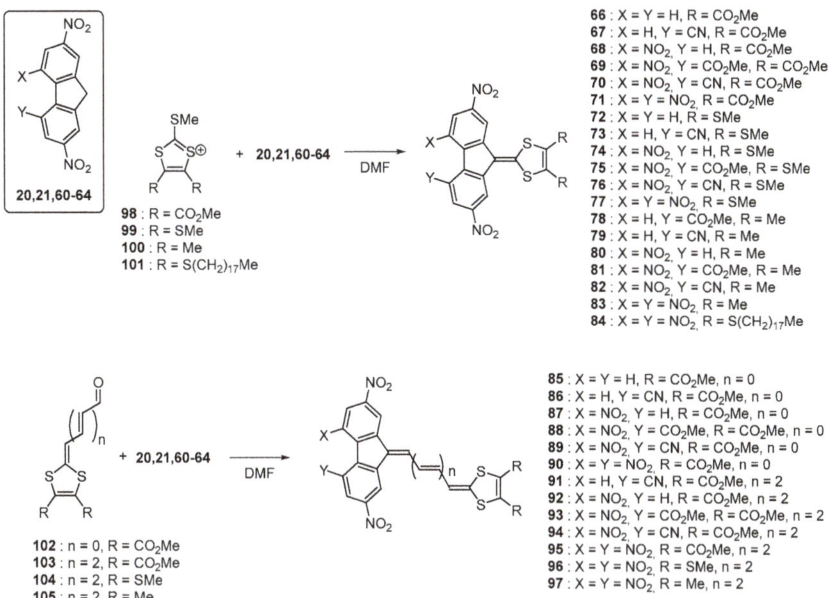

Figure 8. Synthetic routes to push–pull dyes **66–97** based on 1,3-dithiole donor moieties.

Table 3. UV-visible absorption characteristics of compounds **71**, **77**, **83**, **90** and **95–97**.

Compounds	71	77	83	90	95	96	97
ICT Bands (nm) [1]	544	592	611	588	636	707	753
ICT Bands (nm) [2]	535	578	595	572	590	650	700

[1] in 1,2-dichloroethane. [2] in acetone.

As potential application for these structures, the most soluble derivative **84** was examined as a sensitizer in photothermoplastic storage media (PTSM) based on poly[*N*-(2,3-epoxypropyl)carbazole] (PEPC) [44]. Notably, photothermoplastic materials can find numerous applications in aerospace and astrophysics [62]. In the present case, high values of holographic sensitivity in the ICT region were determined, outperforming the results obtained with tetranitrofluorene **21**, when used as the sensitizer for the same polymer. Especially **84** is a promising candidate for this application as it possesses a significant electron affinity (2.1 eV deduced from its reduction potentials (E_{red1} = -0.81 V, E_{red2} = -1.04 V vs. Fc^0/Fc^+ couple), facilitating the formation of charge transfer complexes with electron donors such as poly(carbazoles).

Finally, the best comprise between electron affinity and intramolecular charge transfer energy was obtained with the molecule 106 [63]. This molecule was obtained by condensation of the aldehyde **107** with **21** in DMF (See Figure 9). Comparison of the electrochemical properties of this molecule with those of **21** evidenced the two molecules to exhibit similar redox potentials (-0.89 V, -1.04 V and -1.78 V for 106 vs. -0.88 V, -1.15 V and -1.79 V for 21 vs. Fc^0/Fc^+ couple respectively), providing similar electron affinities and thus electron accepting abilities to the two molecules. However, compared to its analogue **84**, a red-shift of the two ICT bands was obtained for **106**, with a red-shift of 95 and 33 nm for the two ICT bands and an increase of the peak intensity of 1.5 and 3 for the two ICT bands respectively (See Table 4). This enhancement of the molar extinction coefficient is consistent with the elongation of the π-conjugated system, increasing the oscillator strength and thus the electronic delocalisation upon excitation.

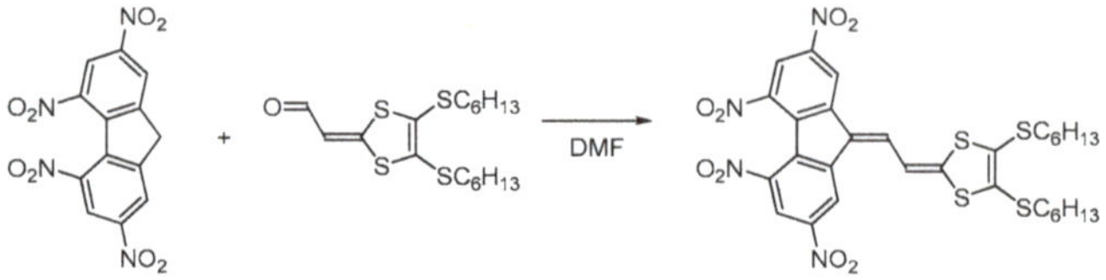

Figure 9. Synthetic routes to tetranitrofluorene (TNF)-based dye **106**.

Table 4. UV-visible absorption characteristics of **84** and **106**.

Compounds	106 [1]	84 [2]
ICT Bands (nm)	545, 643	450, 610

[1] in dichloromethane. [2] in acetone.

While comparing **84** and **106**, the ability of **106** to sensitize the photoconductivity of the carbazole-based polymer PEPK-1 was greatly enhanced, by the red-shift of its absorption bands but also by the similar intensity of the two ICT bands, widening the photoresponse over the 500–700 nm region. Indeed, the photoresponse of **84** was limited to the 550–650 nm region due to the presence of a unique ICT band.

As already mentioned, elongation of the π-conjugated spacer is favorable to increase the molar extinction coefficient of the intramolecular charge transfer and to redshift the absorption maximum. An improvement of the photoconductivity sensitization is also attended if the electron affinity of the chromophore is enhanced, favouring the formation of a charge transfer complex with the sensitized polymer. With aim at optimizing these different parameters (length of the π-conjugated spacer, electron donating/accepting ability of the two partners), a thiophene was introduced between the electron-donating 1,3-dithiole group and the fluorene acceptor (see Figure 10) [64].

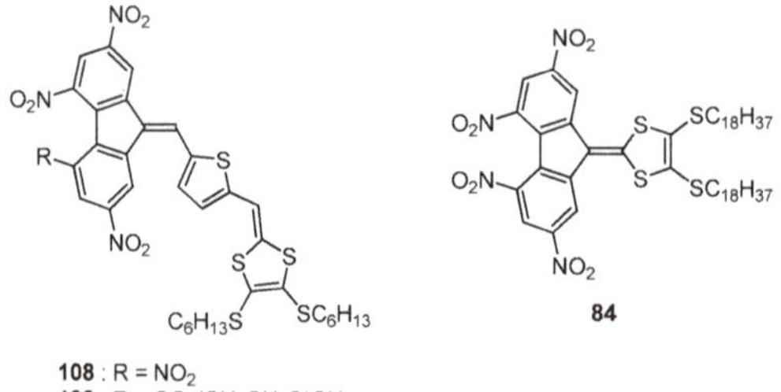

108 : R = NO$_2$
109 : R = CO$_2$(CH$_2$CH$_2$O)CH$_3$

Figure 10. Chemical structures of **108**, **109** and **84**.

From a photophysical point of view, only few details were provided in this work, excepted that the two ICT bands are detected in the 400–900 nm region, with a significant red-shift of the absorption maximum accompanied by a substantial increase of the molar extinction coefficient for **108/109** relative to that of **84**. By electrochemistry, unexpectedly, an irreversible one-electron reduction process was observed for **108** and **109**, identified as resulting from the presence of the hydrogen on the double bond close to the acceptor moiety. As a consequence of this electrochemical irreversibility, when tested as sensitizers for the carbazole-based polymer PEPK, only a poor photoresponse in the ICT band of **108/109** was observed and this result was assigned to the poor photochemical stability of the radical anion previously demonstrated by the irreversibility of the reduction process. Solubility of the sensitizer is another major issue, especially for fluorene derivatives that are prone to aggregate, and this issue was addressed by elongating the chain (**109** vs. **108**). The higher solubility of **109** compared to that of **108** was evidenced.

Recently, the design used for the synthesis of 84 was revisited in a more ambitious study where **21** structures were examined [65]. Their chemical structures are depicted in the Figure 11. In fact, compared to the previous study [61], only two new structures were added, namely **110** and **111**. However, a more detailed study was carried out in this new work.

Figure 11. Synthetic route to **67–84** and **110–111**.

All molecules **67–84** and **110**, **111** exhibited a strong color ranging from red to black and the different dyes were also characterized by a low solubility in most of the common organic solvents. This is the reason **84** was synthesized: in order to greatly improve the solubility of this chromophore. As the main characteristic of this series, the different molecules are characterized by multiple redox states, ranging from four to five redox states what is rarely observed for purely organic molecules [66–68]. Precisely, until four reduction processes could be detected for **71**, **70**, and **68**, what is also observed for fullerene derivatives [69]. Considering that upon reduction, the radical anion, the dianion, the radical trianion, and the tetraanion are centered on the fluorene moiety, a major influence of the substitution pattern was evidenced on these different reduction processes. A summary of the redox potentials is provided in the Table 5. Precisely, a linear correlation for the first and second reduction processes could be determined while using the Hammett constants which quantify the impact of the substitution pattern on the redox potentials [70]. No influence of the substitution pattern of the dithiole moiety on the reduction potentials was detected, evidencing the negative charge to be exclusively localized on the fluorene fragment. These results were confirmed by the theoretical calculations done on the different molecules, evidencing the LUMO level to be mostly centered on the fluorene moiety.

Table 5. Summary of the electrochemical data in dimethylacetamide (DMA) (vs. Fc^0/Fc^+ couple) and optical properties of the different dyes recorded in CH_2Cl_2.

Compounds	E_{ox1} (V)	E_{red1} (V)	E_{red2} (V)	E_{red3} (V)	E_{red4} (V)	λ_{ICT} (nm)
84	0.86, 1.22 irr [1]	−0.91 [1]	−1.10 [1]	−1.91 [1]	-	590
84	-	−0.80	−1.04	−2.04	-	
83	0.85 irr	−0.83	−1.07	−2.01	-	611
82	0.84 irr	−0.94	−1.15	−2.04	-	589
81	0.82, 1.35 irr [1]	−1.11 [1]	−1.21 [1]	−1.89 [1]	-	580
81	0.79 irr	−0.98	−1.22	−2.05	-	
80	0.76 irr	−1.12	−1.35	−2.02	-	554
79	0.73 irr	−1.18	−1.37	-	-	540
78	-	−1.28	−1.43	-	-	535
77	0.79 irr	−0.78	−1.02	−2.00	-	592
76	0.77 irr	−0.90	−1.11	−2.03	-	572
75	0.75 irr	−0.94	−1.18	−2.05	-	568
74	0.73 irr	−1.08	−1.29	−2.00	-	538
73	0.72 irr	−1.14	−1.32	−1.53 irr	-	529
72	-	-	-	-	-	498
110	-	−0.76	−1.01	−1.99	-	558
111	1.18 irr	−0.77	−1.02	−2.00	-	557
71	-	−0.76	−0.98	−1.79	−2.08	544
70	-	−0.87	−1.06	−1.81	−2.10	528
69	-	-	-	-	-	520
68	-	−1.05	−1.25	−1.83	−2.03	496
67	-	−1.11	−1.31	−1.88	-	488
66	-	-	-	-	-	472

[1] in dichloromethane.

While examining the position of the ICT bands, classical trends were determined such as a clear bathochromic shift of the ICT band while improving the accepting ability of the fluorene moiety or by improving the electron donating ability of the 1,3-dithiole moiety. Here again, a linear correlation with the sum of the Hammett parameters was demonstrated, evidencing this series of molecules to exhibit a classical modification of the ICT band with the strength of the donors and the acceptors. The most redshifted absorption was detected for 83, with an ICT band peaking at 611 nm. Only a weak positive solvatochromism was determined for all molecules by increasing the solvent polarity, the variation of the ICT band being lower than 20 nm. This result is indicative of a highly polar ground state. Conversely, a pronounced thermochromic behavior was evidenced with a hypsochromic shift of the ICT band while increasing the temperature. This phenomenon was notably demonstrated in high-boiling point solvents such as chlorobenzene and 1,2-dichlorobenzene. Such a behavior is not so unusual since it was previously reported for polythiophenes [71,72]. Finally, among the series of 27 molecules, only **84** was examined as an electron acceptor for the preparation of CTC with carbazole-containing polymers. Precisely, 84 was used for recording holograms on photothermoplastic storage media (PTSM) materials in combination with poly(*N*-epoxypropylcarbazole) (PEPK). To reach the maximum sensitivity, a concentration of only 2 wt.% was required for **84** compared to 10 wt.% for benchmark sensitizers such as 2,4,7-trinitrofluorenone and 2,4,7-trinitro-9-dicyanomethylene-fluorene, as a result of a broader UV-visible absorption spectrum.

Until 2008, the unique application reported for poly(nitro)fluorene dyes concerned the electrophotographic processes with the sensitization of carbazole-containing polymers [73]. In 2009, the scope of applicability of poly(nitro)fluorene dyes was extended to aggregation-induced emission (AIE) and the design of red-emitting materials [74]. For this purpose, three molecules (**112–114**) were examined and the slight differences in chemical structures drastically impacted the optical properties and the supramolecular structures (See Figure 12). To examine this last point, symmetrically and asymmetrically substituted push–pull molecules were investigated.

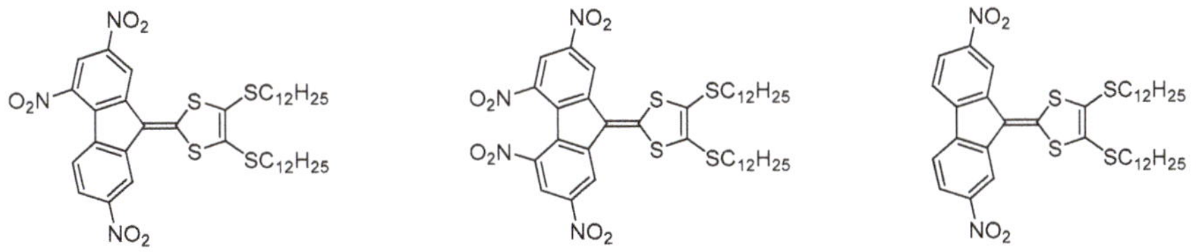

Figure 12. Chemical structures of **112–114**.

While refluxing **112** and **113** in hexanes and cooling the solutions at 0 °C afforded in the two cases precipitates that were examined by optical microscopy. If the formation of spheres of 6–8 μm diameter and a nanostructured surface was observed for **112**, conversely, microtubes of 2.5–5 μm diameter were obtained for **113**. The size and morphology of the microstructures could be easily tuned by varying the cooling temperature. Thus, "flowers with petals" could be obtained while cooling the solution at 15 °C for **112**, replacing the microspheres previously obtained when the solution was cooled at 0 °C. By optically characterizing the microstructures, the two charge transfer bands detected in hexanes at 417 and 520 nm for **112** gradually evolved to a band at 480 nm with a shoulder at 520 nm upon aggregation. This new band was assigned to an intermolecular charge transfer between adjacent molecules in the solid state. Based on this finding, the cohesion between molecules in the solid-state results from intermolecular interactions between molecules, outperforming the intramolecular interactions. A similar behavior was evidenced for **113**, the intramolecular charge transfer band at 552 nm splitting into two new bands at 385 and 460 nm. While examining the photoluminescence properties, a 50-fold enhancement of the photoluminescence quantum yield (4.5%) was determined for **112** upon aggregation. An AIE process was thus evidenced. Conversely, the symmetrical substituted **113** did not show any enhancement of the photoluminescence properties upon aggregation. In fact, authors determined from the crystallographic investigations that the structure of 113 was more twisted than that of **112** in the solid state, quenching the fluorescence. Parallel to this, intermolecular interactions are responsible of the aggregation of **112** in the solid state whereas dipole interactions govern the packing of **113** in the solid state. It was thus concluded the molecular dipole moments in chromophores to be a crucial parameter governing the packing mode and the ability to design AIE emitters.

Parallel to 1,3-dithiole electron donors, thiophenes are also extensively used for the design of semiconducting polymers [75] and materials for NLO applications due to their exceptional electron donating ability and their oxidation potentials [76]. In this context, a series of poly(nitrofluorenes) where the thiophene group was not used as a spacer but as an electron donor were prepared [77]. More precisely, the 1,3-dithiole moiety was fused with the thiophene ring, furnishing an extended donor (see Figure 13).

For this series of dyes **114–131**, the synthesis of the acceptors and especially the selectivity during the mono- and dinitration reactions of the fluorene esters proved to be challenging and the reaction conditions had to be carefully optimized for each of them. To illustrate this, **138–142** could only be obtained by using a HNO_3: AcOH 3:20 ratio starting from **133–137** whereas **143** and **144** could be more easily prepared while using a HNO_3: AcOH 1:1 ratio. Conversely, **148–150** could only be obtained while reacting **133–135** in fuming nitric acid. In the case on long-chain esters (undecanyl and triethyleneglycol monomethyl ether and triethyleneglycol monoethyl ether), the purification of **145** and **146** was almost impossible due to the presence of a mixture of di- and trinitro derivatives that could not be separated due to similar polarities on the column chromatography. As an alternative strategy to access to **145** and **146**, the ester **144** was hydrolyzed in acidic conditions providing **153** and esterified with the appropriate alcohol furnishing the two expected esters **145**, **146** and even **154**.

114 : W = H, X = H, Y = H, Z = H
115 : W = NO₂, X = H, Y = H, Z = H
116 : W = NO₂, X = H, Y = CO₂Bu, Z = H
117 : W = NO₂, X = H, Y = H, Z = NO₂
118 : W = NO₂, X = CO₂Me, Y = H, Z = NO₂
119 : W = NO₂, X = CO₂Bu, Y = H, Z = NO₂
120 : W = NO₂, X = CO₂C₁₁H₂₃, Y = H, Z = NO₂
121 : W = NO₂, X = CO₂(CH₂CH₂O)Et, Y = H, Z = NO₂
122 : W = NO₂, X =NO₂, Y = H, Z = NO₂
123 : W = NO₂, X = CO₂Me, Y = NO₂, Z = NO₂
124 : W = NO₂, X = CO₂Bu, Y = NO₂, Z = NO₂
125 : W = NO₂, X = CO₂(CH₂CH₂O)Me, Y = NO₂, Z = NO₂
126 : W = NO₂, X = NO₂, Y = NO₂, Z = NO₂

127 : W = NO₂, X = H, Y = CO₂Bu, Z = NO₂
128 : W = NO₂, X = CO₂Bu, Y = H, Z = NO₂
129 : W = NO₂, X =NO₂, Y = H, Z = NO₂
130 : W = NO₂, X = CO₂Bu, Y = NO₂, Z = NO₂
131 : W = NO₂, X = NO₂, Y = NO₂, Z = NO₂

Figure 13. Structures of the different push–pull chromophores comprising a fused thiophene-1,3-dithiole **114–126** and dtetrahydrothiophene moieties **127–131**.

As other surprising result, the synthesis of **151** could not be realized in the conditions used for **150**, the Nuclear Magnetic Resonance (NMR) analysis evidencing that the terminal ethyl group was lost (see Figure 14). Loss of the ethyl group and formation of a nitrate was confirmed by Infrared spectroscopy analyses. This unexpected behavior was not observed for **150**.

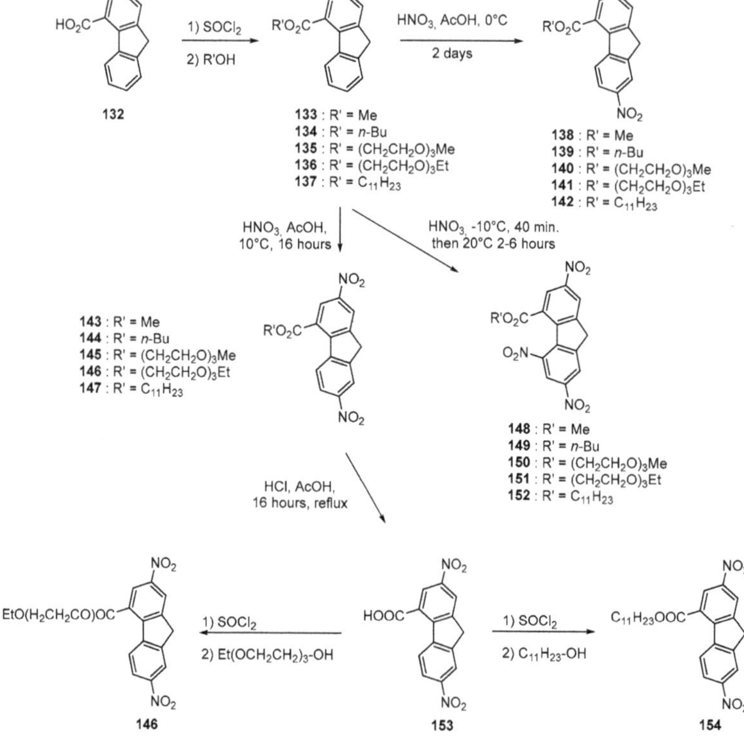

Figure 14. Synthetic routes to the various electron acceptors **132–154**.

From a synthetic point of view, **114–126** were obtained in conditions adapted for each acceptor. Thus, due to the high reactivity of the tri- and tetranitrofluorene acceptors, condensation with salt **155** could be realized simply in DMF whereas pyridine was used as the solvent for the less reactive dinitro acceptors. Long reaction times were required for the reaction with mononitro derivatives, yielding the targeted molecules in low yields (40 to 76% yield) due to the degradation of the salt **155** over time. Compounds **114–116** could not be isolated pure. Face to this result, reactivity of the dihydrothiophene salt **157** towards condensation with fluorene acceptors was examined. Due to a greater delocalisation of the positive charge over the whole salt, reaction temperature for the synthesis of **127–131** could be considerably reduced (room temperature for the synthesis of **127–131** versus 100 °C for **114–125**, 50 °C for **126**) (see Figure 15). Reaction yields ranging from 47 to 64% were determined for the synthesis of **127–131**.

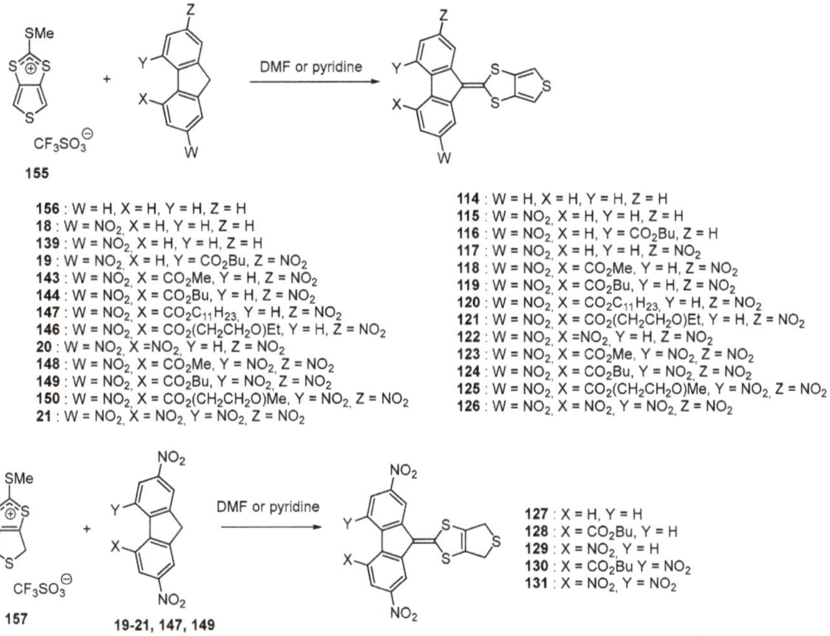
SMe
CF3SO3
155
DMF or pyridine
156 : W = H, X = H, Y = H, Z = H
18 : W = NO2, X = H, Y = H, Z = H
139 : W = NO2, X = H, Y = H, Z = H
19 : W = NO2, X = H, Y = CO2Bu, Z = NO2
143 : W = NO2, X = CO2Me, Y = H, Z = NO2
144 : W = NO2, X = CO2Bu, Y = H, Z = NO2
147 : W = NO2, X = CO2C11H23, Y = H, Z = NO2
146 : W = NO2, X = CO2(CH2CH2O)Et, Y = H, Z = NO2
20 : W = NO2, X =NO2, Y = H, Z = NO2
148 : W = NO2, X = CO2Me, Y = NO2, Z = NO2
149 : W = NO2, X = CO2Bu, Y = NO2, Z = NO2
150 : W = NO2, X = CO2(CH2CH2O)Me, Y = NO2, Z = NO2
21 : W = NO2, X = NO2, Y = NO2, Z = NO2
114 : W = H, X = H, Y = H, Z = H
115 : W = NO2, X = H, Y = H, Z = H
116 : W = NO2, X = H, Y = CO2Bu, Z = H
117 : W = NO2, X = H, Y = H, Z = NO2
118 : W = NO2, X = CO2Me, Y = H, Z = NO2
119 : W = NO2, X = CO2Bu, Y = H, Z = NO2
120 : W = NO2, X = CO2C11H23, Y = H, Z = NO2
121 : W = NO2, X = CO2(CH2CH2O)Et, Y = H, Z = NO2
122 : W = NO2, X =NO2, Y = H, Z = NO2
123 : W = NO2, X = CO2Me, Y = NO2, Z = NO2
124 : W = NO2, X = CO2Bu, Y = NO2, Z = NO2
125 : W = NO2, X = CO2(CH2CH2O)Me, Y = NO2, Z = NO2
126 : W = NO2, X = NO2, Y = NO2, Z = NO2
SMe
CF3SO3
157
NO2
NO2
19-21, 147, 149
DMF or pyridine
NO2
NO2
127 : X = H, Y = H
128 : X = CO2Bu, Y = H
129 : X = NO2, Y = H
130 : X = CO2Bu Y = NO2
131 : X = NO2, Y = NO2

Figure 15. Synthetic route to the series of push–pull molecules **114–131**.

While examining the optical properties of the series of dyes **114–131**, a bathochromic shift of the two ICT bands was unexpectedly observed for **127–131** compared to that of their analogues **114–126**. This counter-intuitive behavior can be assigned to the fact that when the thiophene ring is fused to 1,3-dithiole, this latter doesn't behave as a donor but as an acceptor, reducing the electron donating ability of the 1,3-dithiole moiety. A similar behavior was previously reported for TTF derivatives [78,79]. The most red-shifted ICT was detected for **131**, bearing **21** as the acceptor and the ICT was detected at 596 nm (see Table 6).

Table 6. UV-visible absorption characteristics of push–pull molecules **115–131**.

Compounds	115	116	117	118	119	120
ICT Bands (nm) [1]	429	446	474	492	492	492
Compounds	**121**	**122**	**123**	**124**	**125**	**126**
ICT Bands (nm) [1]	492	508	538	542	540	562
Compounds	**127**	**128**	**129**	**130**	**131**	
ICT Bands (nm) [1]	502	520	537	574	596	

[1] in DMF.

The electrochemical behavior of **114–131** revealed an irreversible oxidation process to take place upon oxidation. As deduced by UV-visible spectroscopy, the negative impact of the electron-withdrawing ability of the fused thiophene on the redox properties was confirmed, the oxidation potentials being detected at more cathodic potentials for the **114–126** vs. **127–131**. An oxidation process centered on the 1,3-dithiole fragment could be confirmed. Two closely-spaced reduction processes could be detected for all molecules with the presence of an additional reduction peak detected around -1.8 V (vs. Fc^0/Fc^+ couple), that was reversible, quasi-reversible or irreversible depending of the electron-withdrawing acceptor. No regular dependence with the substitution pattern of the fluorene moiety was evidenced for this third and additional peak. Finally, attempts to electropolymerize the series **114–126** failed, as a result of the formation of the radical cation onto the 1,3-dithiole part which is thus inappropriate for dimerization. The steric hindrance of the monomer was also suggested as impeding the monomeric units to connect to each other. This drawback was overcome with the terthiophene derivative **158** where two electroactive thiophene groups were introduced on both side of the thieno-1,3-dithiol-2-ylidene moiety (See Figure 16) [80]. In these conditions, a sufficient distance was introduced between the bulky fluorene units so that a polythiophene polymer could be formed by cycling between 0 and 1.40 V.

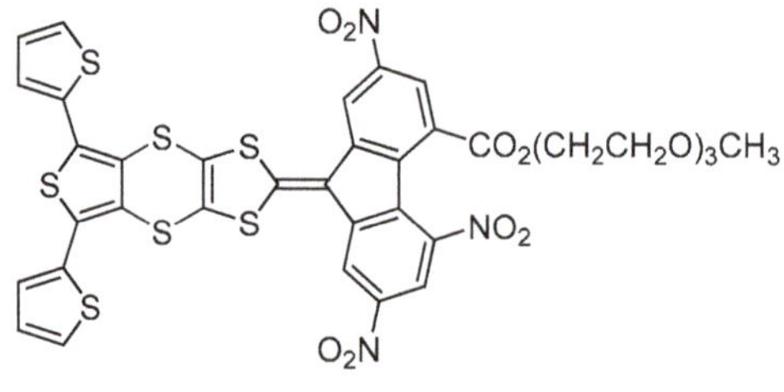

158

Figure 16. Chemical structure of **158**.

To reduce the negative impact of the 1,3-dithiole moiety on the electrochemical oxidation of the thiophene moiety and the localization of the radical cation onto the 1,3-dithiole moiety, a strategy consisting in isolating electronically the two groups from each other was examined and in this aim, the **160–164** series was developed [81]. By analogy with the previous work, a series of molecules **158**, **166–169** where additional and lateral thiophene groups have been introduced to reduce the steric hindrance was also prepared (See Figure 17).

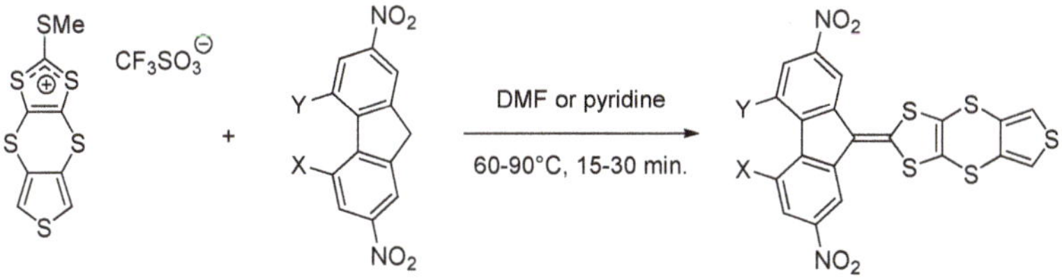

159

19 : X = H, Y = H
146 : X = H, Y = CO$_2$(CH$_2$CH$_2$O)$_3$CH$_2$CH$_3$
20 : X = NO$_2$, Y = H
150 : X = CO$_2$(CH$_2$CH$_2$O)$_3$CH$_3$, Y = NO$_2$
21 : X = NO$_2$, Y = NO$_2$

DMF or pyridine
60-90°C, 15-30 min.

160 : X = H, Y = H
161 : X = H, Y = CO$_2$(CH$_2$CH$_2$O)$_3$CH$_2$CH$_3$
162 : X = NO$_2$, Y = H
163 : X = CO$_2$(CH$_2$CH$_2$O)$_3$CH$_3$, Y = NO$_2$
164 : X = NO$_2$, Y = NO$_2$

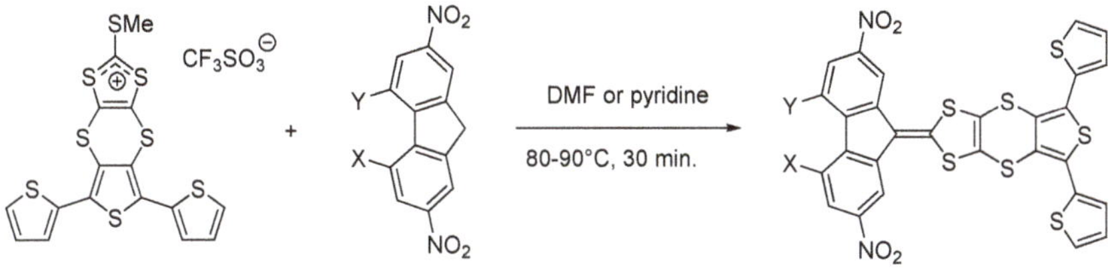

165

19 : X = H, Y = H
146 : X = H, Y = CO$_2$(CH$_2$CH$_2$O)$_3$CH$_2$CH$_3$
20 : X = NO$_2$, Y = H
150 : X = CO$_2$(CH$_2$CH$_2$O)$_3$CH$_3$, Y = NO$_2$
21 : X = NO$_2$, Y = NO$_2$

DMF or pyridine
80-90°C, 30 min.

166 : X = H, Y = H
167 : X = H, Y = CO$_2$(CH$_2$CH$_2$O)$_3$CH$_2$CH$_3$
168 : X = NO$_2$, Y = H
158 : X = CO$_2$(CH$_2$CH$_2$O)$_3$CH$_3$, Y = NO$_2$
169 : X = NO$_2$, Y = NO$_2$

Figure 17. Structures of electropolymerizable monomers **160–164** and **158, 166–169**.

The different molecules were prepared using the standard procedure, by reaction consisting in the condensation of the substituted fluorenes with the appropriate dithiolium salt **159** or **165**. It has to be noticed that **165** is unstable so that it has to be used immediately after synthesis, in the next step, without purification. If **166** and **168** could be synthesized, their low solubilities impeded their purifications so that, even if obtained, these two compounds were thus further investigated.

By UV-visible spectroscopy, the low influence of the thiophene moiety as well as the 1,4-dithiino spacer on the charge transfer interaction was evidenced, especially by comparing the positions of the ICT bands with those of unsubstituted analogues previously reported (See Table 7) [55,61,82] Similarly, comparison of the oxidation potentials of the **167–169** series with those of the series **160–164** previously studied [77] evidenced the 1,4-dithiino spacer to decrease the electronic effects of the fluorene moiety on the oxidation ability of the thiophene unit, what is favorable for the oxidative electropolymerization of thiophene. Finally, for electropolymerization, only **158** was examined, this latter exhibiting the best compromise between solubility and electron-accepting ability of the fluorene fragment.

Table 7. UV-visible absorption characteristics of push–pull molecules **158, 160–164**, and **169**.

Compounds		168		158	169
ICT bands (nm) [1]		503		565	589

Compounds	160	161	162	163	164
ICT bands (nm) [1]	489	513	530	567	588

[1] in DMF.

Poly(158) was obtained by electrochemical oxidation, by successive cycling between 0 and 1.4 V in dichloromethane. A dark blue polymer formed at the surface of the Gold electrode. By UV-visible spectroscopy, an ICT band located at 567 nm could be determined for the electrogenerated polymer. For comparison, poly(158) was also synthesized chemically, by oxidation with $FeCl_3$, and an ICT band at the same position could be detected. By FTIR spectroscopy, analysis of the photoinduced IR spectrum of poly(158) revealed the presence of two representative infrared active vibration (IRAV) bands, one evidencing the presence of radical cations delocalized over the polythiophene backbone and a second one demonstrating the accepting part of the polymer to be localized over the fluorene units.

3. Push–Pull Molecules Obtained by Nucleophilic Substitution

All the aforementioned chromophores have been obtained by functionalization of the methylene group of the fluorene acceptor. In this field, the most popular reaction has undoubtedly been the Knoevenagel reaction. Parallel to the Knoevenagel reaction, an alternative has also been proposed, based on the nucleophilic substitution on the dicyanomethylene fragment of dicyano-methylenefluorene acceptors by aliphatic amines. This reaction has previously been reported for strong electron acceptors such as tetracyanoquinodimethane (TCNQ) and tetracyanoethylene (TCNE) [83–85]. In the present case, one or two cyano groups can be substituted, depending of the reaction conditions. In 1995, Perepichka, et al. reported a series of push–pull molecules **178–193** and **194–209** prepared from **170–177**, based on the successive substitution of a cyano group by a secondary amine (See Figure 18) [86]. In this work, two different amines were investigated, namely piperidine and morpholine.

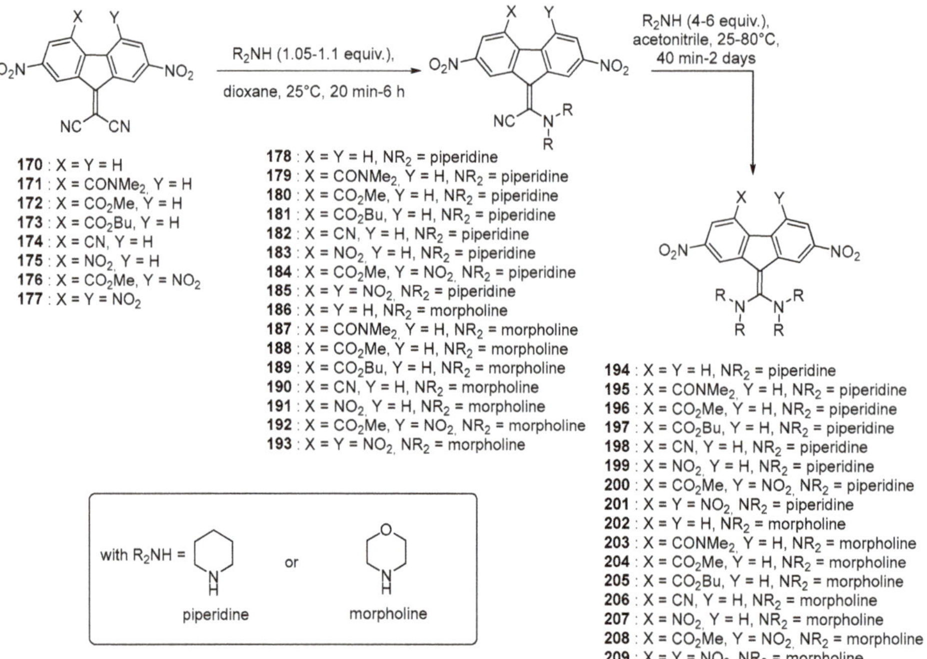

Figure 18. Structures of push–pull molecules **178–193** and **194–209**.

By the simultaneous presence of the electron donating aliphatic amino group and the electron-accepting fluorene fragment within the same molecule, appearance of an intramolecular charge transfer band located in the visible range could be detected. Interestingly, by NMR, a Z/E isomerization could be evidenced and by increasing the number of electron-withdrawing group on the fluorene fragment contributed to lower the rotation barrier. In this work, only few details were provided concerning the UV-visible absorption spectra of **178–193** and **194–209**. The unique informations concern the absorption range, going from 440 to 550 nm for **178–193** (440–550 nm) and 530–620 nm for **194–209** respectively. More efforts were devoted to determine the relationship existing between the position of the ICT band and the substituent effects. Especially, the Koppel–Palm four-parameters' equation taking account from the polarity, the polarizability, the acidity and the basicity of the solvent was examined [56]. In this study, only the polarity and the basicity proved to be relevant to justify the shift with the substitution pattern.

Following this work, a more detailed study was proposed in 1996 by the same authors and the lateral functionalization of the fluorene core was investigated (see Figure 19) [82]. By electrochemistry, the specific behavior of **204** compared to the series **210–217** and **218–225** was evidenced. Notably, a difference of 560 mV was found between the two first reduction waves of **204** whereas this difference is only of 160 mV for **210–217** and **218–225** (see Table 8). The electron affinity (2.46 eV) of **204** was determined as being greater than that of the series **210–217** (1.65–2.03 eV) and the series **218–225** (1.36–1.85 eV). This result is counter-intuitive, considering the presence of the electron-donating group directly connected to the fluorene core but can rationalized on the basis of the sum of the nucleophilic constants. Considering that the two series **210–217** and **218–225** exhibit a strong absorption band in the visible range (see Table 8), the generation of the radical anion or the dianion of these structures should drastically reduce the electron accepting ability of the fluorene moiety, resulting in the disappearance of the intramolecular charge transfer band. A similar behavior should also be observed upon generation of the radical cation, making these structures appealing candidates for electrochromic applications. These conclusions were confirmed by spectro-electrochemical experiments, demonstrating a reduction of the ICT band upon oxidation and/or reduction of the chromophore, associated with the appearance of new bands corresponding to the formation of the radical cation, radical anion or dianion.

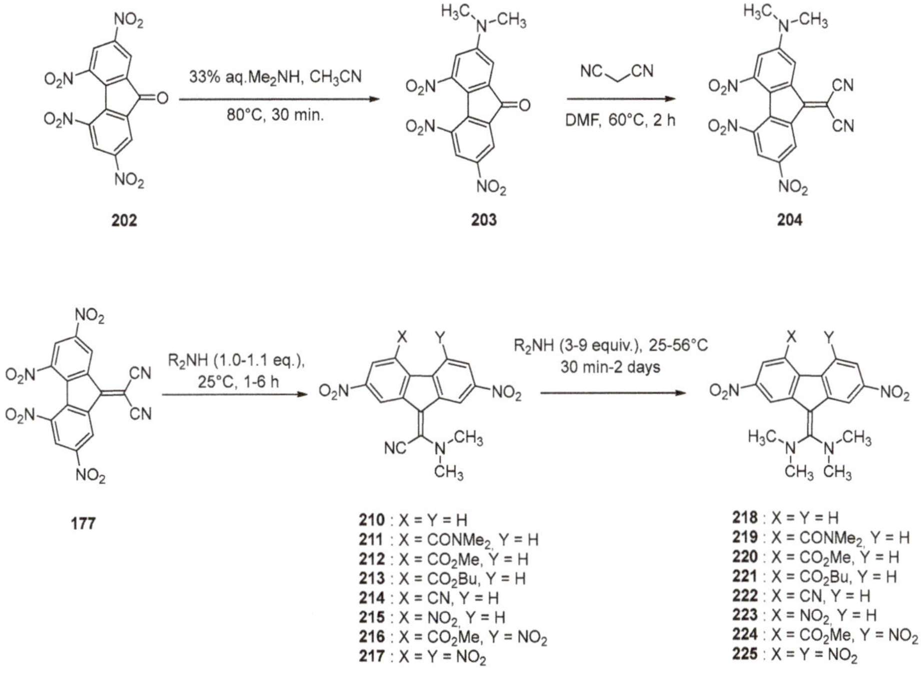

210 : X = Y = H
211 : X = CONMe$_2$, Y = H
212 : X = CO$_2$Me, Y = H
213 : X = CO$_2$Bu, Y = H
214 : X = CN, Y = H
215 : X = NO$_2$, Y = H
216 : X = CO$_2$Me, Y = NO$_2$
217 : X = Y = NO$_2$

218 : X = Y = H
219 : X = CONMe$_2$, Y = H
220 : X = CO$_2$Me, Y = H
221 : X = CO$_2$Bu, Y = H
222 : X = CN, Y = H
223 : X = NO$_2$, Y = H
224 : X = CO$_2$Me, Y = NO$_2$
225 : X = Y = NO$_2$

Figure 19. Structure of push–pull molecule **204** and the two series **210–217**, **218–225**.

Table 8. Summary of the electrochemical data in acetonitrile (vs. Fc^0/Fc^+ couple) and optical properties of the different dyes **210–224** recorded in dioxane.

Compounds	E_{ox1} (V) [1]	E_{red1} (V) [1]	E_{red2} (V) [1]	E_{red3} (V) [1]	E_{red4} (V) [1]	λ_{ICT} (nm) [2]
210	1.23	−0.89	-	-	-	441
211	1.33	−0.82	-	-	-	454
212	1.33	−0.79	-	-	-	460
213	1.34	−077	-	-	-	459
214	1.36	−0.70	−0.77	-	0.07	465
215	1.37	−0.67	−0.75	−1.32	0.08	472
216	1.47	−0.53	−0.62	−1.34	0.09	492
217	1.55	−0.40	−0.50	−1.31	0.10	505
218	0.71	−1.16	-	-	-	528
219	-	-	-	-	-	532
220	0.77	−1.04	−1.14	-	0.10	543
221	0.77	−1.03	1.13	-	0.10	542
222	0.87	−0.97	−1.06	-	0.09	539
223	0.89	−0.93	−1.03	−1.44	0.10	552
224	1.00	−0.84	−0.96	−1.50	0.12	564
225	1.11	−0.70	−0.86	−1.50	0.16	571

[1] in acetonitrile. [2] in dioxane.

Finally, due to the broad absorption of these structures, the sensitization of the photoconductivity of carbazole-containing polymers was examined. While comparing the sensitization ability of **177**, **217** and **225**, a decrease in the order **177** > **217** > **225** was demonstrated, resulting from a reduction of the electron accepting ability of the fluorene core. An hypsochromic shift of the electrophotographic sensitivity was also observed, following the trend of the ICT band.

4. Push–Pull Molecules Based on the Aviram and Ratner Concept

The unimolecular rectifier proposed by Aviram and Ratner in 1974 has excited during numerous years the imagination of researchers [87]. If proposed theoretically, numerous attempts have been carried out to synthesize this conceptual model **226**, without success [88–90]. The difficulty of synthesis lies in the combination within a unique molecule of a strong donor and a strong acceptor that will form an intermolecular charge transfer complex prior to the formation of the covalent bond [91–94]. If the initial concept was based on the covalent linkage of the remarkable electron donor tetrathiafulvalene (TTF) to the strong electron acceptor tetracyanoquinodimethane (TCNQ), numerous acceptors of lower electron-accepting ability were examined, as exemplified by quinones [95,96], tetracyanoanthraquinodimethane [89], fullerenes [48,97]. The conversion of weak acceptors to stronger acceptors subsequently to their covalent linkages has been extensively studied, but this conversion proved to be more difficult than anticipated and most of the attempts failed [88–90]. In 2002, Perepichka, et al. examined the combination of TTF with poly(nitro)fluorene acceptors. Three molecules were synthesized (See Figure 20) [98]. Here again, the strategy consisted in first connecting the fluorenone-based acceptors to the donors, providing the diads **227**, **229**, and **231**. In a second step, **227**, **229**, and **231** were converted to **228**, **230**, and **232** by reaction with malononitrile in DMF at room temperature. Concerning their designs, for **228** and **230**, the length of the flexible spacer was selected to be sufficient to allow an intramolecular electron transfer whereas a short spacer was selected for **232**, precluding the formation of an intramolecular interaction. Due to the use of strong electron donors and acceptors, an intramolecular electron transfer could effectively occur in **228** and **230** thanks to the use of a flexible spacer. The first manifestation of this ICT was the detection by EPR of a strong signal indicative of the formation of a radical determined as being the radical cation of TTF. Conversely, the radical anion of the acceptor was not detected, and this absence of signal was assigned to the formation of head to head dimers, quenching the signal. This behavior has previously been reported for TCNQ derivatives [99]. From an electrochemical point of view, diads **228**, **230** and **232** are characterized by a multiredox behavior, with the formation of the radical cation and dication of TTF and three reversible reduction processes centered on the fluorene moiety corresponding to the formation of the radical anion, dianion and radical trianion (See Table 9). Comparison of the redox potentials of **228**, **230** and the short-chain **232** revealed in this last case, the redox potentials of TTF not to be affected by the presence of the electron acceptor, and this was especially revealed by comparing the oxidation potentials of TTF in **231** and **232** (less than 10 mV of difference). Indeed, in this last case, **232** has been designed to preclude any intramolecular charge transfer and the two partners, in turn, behave as independent molecules. As final interesting feature, an extremely small HOMO-LUMO gap was determined for the three molecules, around 0.3 eV.

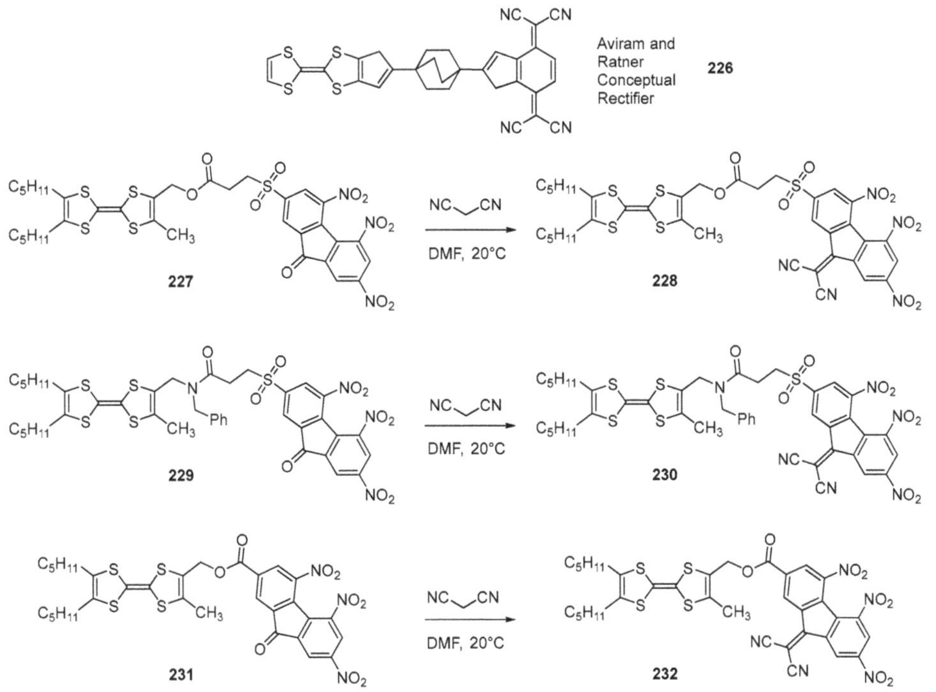

Figure 20. Structures of push–pull chromophores **227–232** derived from the Aviram and Ratner conceptual unimolecular rectifier **226**.

By UV-visible spectroscopy, the six push–pull molecules are characterized by two ICT bands, one in the visible range and another one in the NIR region. As anticipated, improvement of the electron-accepting ability resulted for **228** and **230** to a red-shift of the two absorption bands relative to that of **227** and **229**. Examination of the concentration dependence of the ICT band intensity was determined as being linear for **227**, **229**, **231** whereas a complete disappearance of the charge transfer band was detected for **231** and **232** for concentrations below 10^{-4} M. It can therefore be concluded that the two bands detected for **231** and **232** correspond to through-space interactions between adjacent molecules. Finally, **199** and **201** were examined for their electrochromic properties. **230** proved to be more promising than **228** as a partial decomposition was observed during cycling. This instability was partially assigned to the linker introduced between the donor and the acceptor, the ester connection being less stable than the amide connection. **230** proved to be an interesting material for photochromic applications as the two first oxidation and reduction processes are reversible, ensuring the regeneration of the initial molecule upon cycling.

Table 9. Summary of the electrochemical data in CH_2Cl_2 (vs. Fc^0/Fc^+ couple) and optical properties of the different dyes **198–203** recorded in CH_2Cl_2.

Compounds	E_{ox1} (V)	E_{ox2} (V)	E_{red1} (V)	E_{red2} (V)	E_{red3} (V)	λ_{ICT1} (nm)	λ_{ICT2} (nm)
227	−0.13	0.37	−0.68	−1.04	−1.86	750	990
228	−0.09	0.35	−0.39	−0.90	−1.57	800	1230
229	−0.17	0.31	−0.69	−1.06	−1.84	775	975
230	−0.10	0.32	−0.38	−0.90	−1.56	800	1260
231	−0.11	0.40	−0.72	−1.00	−1.81	630 [1]	900 [1]
232	−0.10	0.41	−0.39	−0.93	−1.62	785–825	1200–1330

[1] in acetone.

5. Intramolecular Charge Transfer Complexes in Metal-Based Dyes

Organic electron donors were not the only ones to be used for the preparation of push–pull chromophores comprising poly(nitro)fluorene acceptors. Ferrocene was also examined as a potential donor and influence of the length as well as the nature of the spacer introduced between the two fragments was studied (See Figure 21) [100].

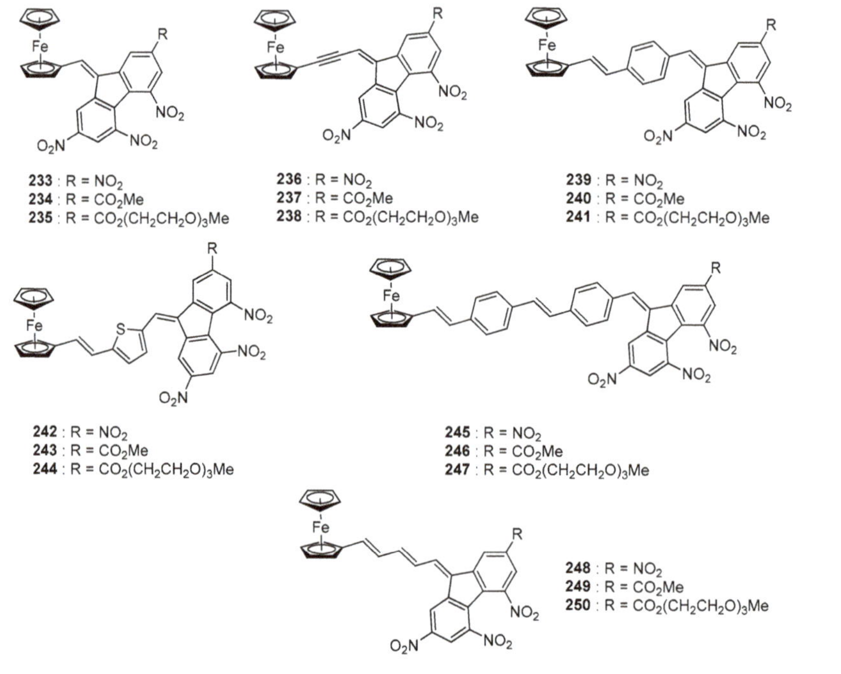

233 : R = NO_2
234 : R = CO_2Me
235 : R = $CO_2(CH_2CH_2O)_3Me$

236 : R = NO_2
237 : R = CO_2Me
238 : R = $CO_2(CH_2CH_2O)_3Me$

239 : R = NO_2
240 : R = CO_2Me
241 : R = $CO_2(CH_2CH_2O)_3Me$

242 : R = NO_2
243 : R = CO_2Me
244 : R = $CO_2(CH_2CH_2O)_3Me$

245 : R = NO_2
246 : R = CO_2Me
247 : R = $CO_2(CH_2CH_2O)_3Me$

248 : R = NO_2
249 : R = CO_2Me
250 : R = $CO_2(CH_2CH_2O)_3Me$

Figure 21. Push–pull chromophores **233–250** based on ferrocene.

For the series of metal complexes **233–250**, an absorption band ranging from 450 to 900 nm could be detected. These different complexes being designed for Non-Linear Optical (NLO) applications and in order to facilitate the determination of the quadratic hyperpolarizability by the electric field induced second harmonic (EFISH) generation, a soluble version of each complex was synthesized, namely, **235**, **238**, **241**, **244**, **247** and **250**. Going from **235** to **250**, the $\mu\beta$ values increased from 100×10^{-48} esu for **235** to 200×10^{-48} esu for **238**, 800×10^{-48} esu for **241** and 2400×10^{-48} esu for **250**. Evolution of the hyperpolarizability is consistent with the elongation of the π-conjugated spacer in these structures. While examining the optical properties, two ICT could be detected for all compounds. Only a minor variation of the maximum absorption was observed with the spacer (See Table 10).

Table 10. Summary of the optical properties of the different dyes recorded in 1,2-dichloroethane.

Compounds	233	234	236	238	239	241
λ_{ICT1} (nm)	430	410	451	431	430	430
λ_{ICT2} (nm)	620	609	658	616	600	600

Compounds	244	247	248	249	250
λ_{ICT1} (nm)	502	450	464, 514	450, 485	455, 490
λ_{ICT2} (nm)	640	-	706	660	660

Finally, in 2001, the same authors investigated the design of A-D-A triads **233**, **251–258** where ferrocene or a ruthenium complex was used as the donor (See Figure 22) [101]. In these series, the higher electron donating ability of ferrocene compared to the ruthenium complex was demonstrated, a red-shift of the ICT band being detected for **253** (622 nm in 1,2-dichloroethane) compared to **256** (521 nm in 1,2-dichloroethane). Comparison of the UV-visible spectra of the triads **253–255** with the diads **233**, **251**, **252** revealed the position of the ICT bands to remain almost unchanged. Face to these considerations and considering the difficulty to synthesize the triads, it can be concluded that the synthesis of the diads is sufficient and that the connection of the donor to an additional electron acceptor can't improve its electron-donating ability irrespective of the electron-withdrawing ability of the acceptor.

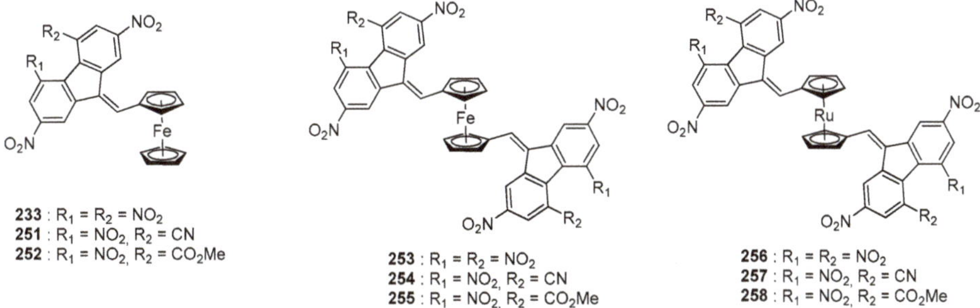

Figure 22. Push–pull chromophores based on metallocenes (iron and ruthenium complexes).

6. Conclusions

In this review, an overview of the different dyes designed with the fluorene scaffold substituted with three or four nitro groups is presented. As detailed in this review, organic dyes have focused the main interest of researchers, but metal-based chromophores have also been investigated. If poly(nitro)fluorene acceptors constitute a unique class of acceptor by their inherent absorption in the near infrared region, the presence of numerous nitro group drastically limits the solubility of the resulting push–pull dyes and numerous efforts have been devoted to overcome this drawback. At present, the scope of applicability of these structures is quite limited, since the compounds have mostly been investigated as sensitizers for hologram recording, and more scarcely for NLO or electrochromic applications. With regards to the recent interest of compounds absorbing in the near infrared region (photopolymerization, dyes for photovoltaics applications, etc.), clearly, the number of applications involving the use of poly(nitro)fluorene structures will greatly expand in the future.

Author Contributions: Writing-Original Draft Preparation, G.N. and F.D.; Writing-Review & Editing, G.N. and F.D.

Funding: This research was funded by Aix Marseille University and the Centre National de la Recherche Scientifique (CNRS).

Conflicts of Interest: The authors declare no conflict of interest.

References

1. Kirtley, J.R.; Mannhart, J. Organic electronics: When TTF met TCNQ. *Nat. Mater.* **2008**, *7*, 520–521. [CrossRef] [PubMed]
2. Jérôme, D.; Mazaud, A.; Ribault, M.; Bechgaard, K. Superconductivity in a synthetic organic conductor (TMTSF)$_2$PF$_6$. *J. Phys. Lett.* **1980**, *41*, 95–98. [CrossRef]
3. Hagfeldt, A.; Boschloo, G.; Sun, L.; Kloo, L.; Pettersson, H. Dye-Sensitized Solar Cells. *Chem. Rev.* **2010**, *110*, 6595–6663. [CrossRef] [PubMed]
4. Holmgaard List, N.; Zalesńy, R.; Arul Murugan, N.; Kongsted, J.; Bartkowiak, W.; Ågren, H. Relation between Nonlinear Optical Properties of Push–Pull Molecules and Metric of Charge Transfer Excitations. *J. Chem. Theory Comput.* **2015**, *11*, 4182–4188. [CrossRef] [PubMed]
5. Verbiest, T.; Houbrechts, S.; Kauranen, M.; Clays, K.; Persoons, A. Second-order nonlinear optical materials: Recent advances in chromophore design. *J. Mater. Chem.* **1997**, *7*, 2175–2189. [CrossRef]
6. Reichardt, C. *Solvents and Solvent Effects in Organic Chemistry*, 3rd ed.; Wiley-VCH: Weinheim, Germany, 2004.
7. Bures, F.; Pytela, O.; Kivala, M.; Diederich, F. Solvatochromism as an efficient tool to study *N,N*-dimethylamino- and cyano-substituted π-conjugated molecules with an intramolecular charge-transfer absorption. *J. Phys. Org. Chem.* **2011**, *24*, 274–281. [CrossRef]
8. Bures, F.; Pytela, O.; Diederich, F. Solvent effects on electronic absorption spectra of donor-substituted 11,11,12,12-tetracyano-9, 10-anthraquinodimethanes (TCAQs). *J. Phys. Org. Chem.* **2009**, *22*, 155–162. [CrossRef]
9. Raymo, F.M.; Tomasulo, M. Optical Processing with Photochromic Switches. *Chem. Eur. J.* **2006**, *12*, 3186–3193. [CrossRef] [PubMed]
10. Martin, N.; Seoane, C. *Handbook of Organic Conductive Molecules and Polymers*; Nalwa, H.S., Ed.; Wiley: Chichester, UK, 1997; Volume 1, pp. 1–86.
11. Martin, N.; Segura, J.L.; Seoane, C. Design and synthesis of TCNQ and DCNQI type electron acceptormolecules as precursors for 'organic metals'. *J. Mater. Chem.* **1997**, *7*, 1661–1676. [CrossRef]
12. Tsubata, Y.; Suzuki, T.; Miyashi, T. Single-component organic conductors based on neutral radicals containing the pyrazino-TCNQ skeleton. *J. Org. Chem.* **1992**, *57*, 6749–6755. [CrossRef]
13. Sullivan, P.A.; Dalton, L.R. Theory-Inspired Development of Organic Electro-optic Materials. *Acc. Chem. Res.* **2010**, *43*, 10–18. [CrossRef] [PubMed]
14. Stahelin, M.; Zysset, B.; Ahlheim, M.; Marder, S.R.; Bedworth, P.V.; Runser, C.; Barzoukas, M.; Fort, A. Nonlinear optical properties of push–pull polyenes for electro-optics. *J. Opt. Soc. Am. B* **1996**, *13*, 2401–2407. [CrossRef]
15. Chen, F.; Zhang, J.; Wan, X. Design and Synthesis of Piezochromic Materials Based on Push–Pull Chromophores: A Mechanistic Perspective. *Chem. Eur. J.* **2012**, *18*, 4558–4567. [CrossRef] [PubMed]
16. Bures, F. Fundamental aspects of property tuning in push–pull molecules. *RSC Adv.* **2014**, *4*, 58826–58851. [CrossRef]
17. Blanchard-Desce, M.; Wortmann, R.; Lebus, S.; Lehn, J.-M.; Krämer, P. Intramolecular charge transfer in elongated donor-acceptor conjugated polyenes. *Chem. Phys. Lett.* **1995**, *243*, 526–532. [CrossRef]
18. Rancic, M.P.; Stojiljkovic, I.; Milosevic, M.; Prlainovic, N.; Jovanovic, M.; Milcic, M.K.; Marinkovic, A.D. Solvent and substituent effect on intramolecular charge transfer in 5-arylidene-3-substituted- 2,4-thiazolidinediones: Experimental and theoretical study. *Arabian J. Chem.* **2017**.
19. Moreno-Yruela, C.; Garín, J.; Orduna, J.; Franco, S.; Quintero, E.; López Navarrete, J.T.; Diosdado, B.E.; Villacampa, B.; Casado, J.; Andreu, R. D−π–A Compounds with Tunable Intramolecular Charge Transfer Achieved by Incorporation of Butenolide Nitriles as Acceptor Moieties. *J. Org. Chem.* **2015**, *80*, 12115–12128. [CrossRef] [PubMed]
20. Guerlin, A.; Dumur, F.; Dumas, E.; Miomandre, F.; Wantz, G.; Mayer, C.R. Tunable Optical Properties of Chromophores Derived from Oligo(p-phenylene vinylene). *Org. Lett.* **2010**, *12*, 2382–2385. [CrossRef] [PubMed]
21. Tehfe, M.-A.; Dumur, F.; Graff, B.; Gigmes, D.; Fouassier, J.-P.; Lalevée, J. Blue-to-Red Light Sensitive Push–Pull Structured Photoinitiators: Indanedione Derivatives for Radical and Cationic Photopolymerization Reactions. *Macromolecules* **2013**, *46*, 3332–3341. [CrossRef]
22. Tehfe, M.-A.; Dumur, F.; Graff, B.; Morlet-Savary, F.; Gigmes, D.; Fouassier, J.-P.; Lalevée, J. Push–pull (thio)barbituric acid derivatives in dye photosensitized radical and cationic polymerization reactions under 457/473 nm laser beams or blue LEDs. *Polym. Chem.* **2013**, *4*, 3866–3875. [CrossRef]

23. Klikar, M.; Jelínková, V.; Růžičková, Z.; Mikysek, T.; Pytela, O.; Ludwig, M.; Bureš, F. Malonic Acid Derivatives on Duty as Electron-Withdrawing Units in Push–Pull Molecules. *Eur. J. Org. Chem.* **2017**, 2764–2779. [CrossRef]

24. Dumur, F.; Mayer, C.R.; Dumas, E.; Miomandre, F.; Frigoli, M.; Sécheresse, F. New Chelating Stilbazonium-Like Dyes from Michler's Ketone. *Org. Lett.* **2008**, *10*, 321–324. [CrossRef] [PubMed]

25. Khattab, T.A.; Gaffer, H.E. Synthesis and application of novel tricyanofuran hydrazone dyes as sensors for detection of microbes. *Color. Technol.* **2016**, *132*, 460–465. [CrossRef]

26. Davis, M.C.; Groshens, T.J.; Parrish, D.A. Preparation of Cyan Dyes from 6-Diethylaminobenzo[*b*]furan-2-carboxaldehyde. *Synth. Commun.* **2010**, *1*, 3008–3020. [CrossRef]

27. Andreu, R.; Carrasquer, L.; Franco, S.; Garın, J.; Orduna, J.; de Baroja, N.M.; Alicante, R.; Villacampa, B.; Allain, M. 4*H*-Pyran-4-ylidenes: Strong Proaromatic Donors for Organic Nonlinear Optical Chromophores. *J. Org. Chem.* **2009**, *74*, 6647–6657. [CrossRef] [PubMed]

28. Andreu, R.; Galan, E.; Garin, J.; Herrero, V.; Lacarra, E.; Orduna, J.; Alicante, R.; Villacampa, B. Linear and V-Shaped Nonlinear Optical Chromophores with Multiple 4*H*-Pyran-4-ylidene Moieties. *J. Org. Chem.* **2010**, *75*, 1684–1692. [CrossRef] [PubMed]

29. Cho, B.R.; Son, K.H.; Lee, S.H.; Song, Y.S.; Lee, Y.K.; Jeon, S.J.; Choi, J.H.; Lee, H.; Cho, M. Two Photon Absorption Properties of 1,3,5-Tricyano-2,4,6-tris(styryl)benzene Derivatives. *J. Am. Chem. Soc.* **2001**, *123*, 10039–10045. [CrossRef] [PubMed]

30. Zhang, X.-H.; Zhan, Y.-H.; Chen, D.; Wang, F.; Wang, L.-Y. Merocyanine dyes containing an isoxazolone nucleus: Synthesis, X-ray crystal structures, spectroscopic properties and DFT studies. *Dyes Pigm.* **2012**, *93*, 1408–1415. [CrossRef]

31. Marinado, T.; Hagberg, D.P.; Hedlund, M.; Edvinsson, T.; Johansson, E.M.J.; Boschloo, G.; Rensmo, H.; Brinck, T.; Sun, L.; Hagfeldty, A. Rhodanine dyes for dye-sensitized solar cells: Spectroscopy, energy levels and photovoltaic performance. *Phys. Chem. Chem. Phys.* **2009**, *11*, 133–141. [CrossRef] [PubMed]

32. Hirosawa, R.; Akune, Y.; Endo, N.; Hatano, S.; Hosokai, T.; Sato, H.; Matsumoto, S. A variety of solid-state fluorescence properties of pyrazine dyes depending on terminal substituents. *Dyes Pigm.* **2017**, *146*, 576–581. [CrossRef]

33. Kulhanek, J.; Bures, F.; Wojciechowski, A.; Makowska-Janusik, M.; Gondek, E.; Kityk, I.V. Optical operation by chromophores featuring 4,5-dicyanoimidazole embedded within poly(methyl methacrylate) matrices. *J. Phys. Chem. A* **2010**, *114*, 9440–9446. [CrossRef] [PubMed]

34. Ci, Z.; Yu, X.; Bao, M.; Wang, C.; Ma, T. Influence of the benzo[*d*]thiazole-derived π-bridges on the optical and photovoltaic performance of D–π–A dyes. *Dyes Pigm.* **2013**, *96*, 619–625. [CrossRef]

35. Bodedla, G.B.; Justin Thomas, K.R.; Fan, M.S.; Ho, K.C. Bi-anchoring Organic Dyes that Contain Benzimidazole Branches for Dye-Sensitized Solar Cells: Effects of π Spacer and Peripheral Donor Groups. *Chem. Asian J.* **2016**, *11*, 2564–2577. [CrossRef] [PubMed]

36. Baheti, A.; Justin Thomas, K.R.; Lee, C.-P.; Li, C.-T.; Ho, K.-C. Organic dyes containing fluoren-9-ylidene chromophores for efficient dye-sensitized solar cells. *J. Mater. Chem. A* **2014**, *2*, 5766–5779. [CrossRef]

37. Raimundo, J.-M.; Blanchard, P.; Gallego-Planas, N.; Mercier, N.; Ledoux-Rak, I.; Hierle, R.; Roncali, J. Design and Synthesis of Push−Pull Chromophores for Second-Order Nonlinear Optics Derived from Rigidified Thiophene-Based π-Conjugating Spacers. *J. Org. Chem.* **2002**, *67*, 205–218. [CrossRef] [PubMed]

38. Levinskii, S.S.; Khitrovo, I.A.; Krivosheeva, L.V. Unified Procedure for the Determination of Polycyclic Aromatic Hydrocarbons and Their Nitro Derivatives by Low-Temperature Luminescence. *J. Anal. Chem.* **2001**, *56*, 1098–1104. [CrossRef]

39. Browning, C.; Hudson, J.M.; Reinheimer, E.W.; Kuo, F.L.; McDougald, R.N., Jr.; Rabaa, H.; Pan, H.; Bacsa, J.; Wang, X.; Dunbar, K.R.; Shepherd, N.D.; Omary, M.A. Synthesis, Spectroscopic Properties, and Photoconductivity of Black Absorbers Consisting of Pt(Bipyridine)(Dithiolate) Charge Transfer Complexes in the Presence and Absence of Nitrofluorenone Acceptors. *J. Am. Chem. Soc.* **2014**, *136*, 16185–16200. [CrossRef] [PubMed]

40. Ong, B.S. Synthesis, intramolecular charge-transfer interaction and electron transport properties of 9,9-bis(alkylthio)nitrofluorenes. *J. Chem. Soc. Chem. Commun.* **1984**, 266–268. [CrossRef]

41. Ong, B.S. A simple and efficient method of thioacetal - and ketalization. *Tetrahedron Lett.* **1980**, *21*, 4225–4228. [CrossRef]

42. Hoegl, H.; Barchietto, G.; Tar, D. Photosensitization of poly(N-vinyl)carbazole. *Photochem. Photobiol.* **1972**, *16*, 335–352. [CrossRef]

43. Enomoto, T.; Hatano, M. Charge-transfer complexes of poly(N-vinylcarbazole) with organic electron acceptors. *Die Makromol. Chem.* **1974**, *175*, 57–65. [CrossRef]

44. Perepichka, I.F.; Mysyk, D.D.; Sokolov, N.I. *Current Trends in Polymer Photochemistry*; Allen, N.S., Edge, M., Bellobono, I.R., Selli, E., Eds.; Ellis Horwood: New York, NY, USA, 1995; p. 318.

45. Mort, J.; Emerald, R.L. Energy-level structure and carrier transport in the poly(N-vinyl carbazole): Trinitrofluorenone charge-transfer complex. *J. Appl. Phys.* **1974**, *45*, 175–178. [CrossRef]

46. Emerald, R.L.; Mort, J. Transient photoinjection of electrons from amorphous selenium into trinitrofluorenone. *J. Appl. Phys.* **1974**, *45*, 3943–3945. [CrossRef]

47. Pfister, G.; Grammatica, S.; Mort, J. Trap-Controlled Dispersive Hopping Transport. *Phys. Rev. Lett.* **1976**, *37*, 1360–1363. [CrossRef]

48. Kuder, J.E.; Pochan, J.M.; Turner, S.R.; Hinrnan, D.F. Fluorenone derivatives as electron transport materials: The relationship of electron affinity and electrochemistry with photoelectric behavior. *J. Electrochem. Soc.* **1978**, *125*, 1750–1758. [CrossRef]

49. Mysyk, D.D.; Perepichka, I.F. Fluorene Compounds with Intramolecular Charge Transfer Containing Dithiolylidene and Selenathiolylidene Substituents. *Phosphorus Sulfur* **1994**, *95*, 527–529. [CrossRef]

50. Bill, N.L.; Ishida, M.; Bähring, S.; Lim, J.M.; Lee, S.; Davis, C.M.; Lynch, V.M.; Nielsen, K.A.; Jeppesen, J.O.; Ohkubo, K.; Fukuzumi, S.; Kim, D.; Sessler, J.L. Porphyrins Fused with Strongly Electron-Donating 1,3-Dithiol-2-ylidene Moieties: Redox Control by Metal Cation Complexation and Anion Binding. *J. Am. Chem. Soc.* **2013**, *135*, 10852–10862. [CrossRef] [PubMed]

51. Broggi, J.; Terme, T.; Vanelle, P. Organic Electron Donors as Powerful Single-Electron Reducing Agents in Organic Synthesis. *Angew. Chem. Int. Ed.* **2014**, *53*, 384–413. [CrossRef] [PubMed]

52. Haley, N.F.; Fichtner, M.W. Efficient and general synthesis of 1,3-dithiole-2-thiones. *J. Org. Chem.* **1980**, *45*, 175–177. [CrossRef]

53. Shafie, A.; Lalezari, I.; Savabi, F. A Facile Synthesis of 5-Substituted 2-Thioxo-1,3-thiaselenoles. *Synthesis* **1977**, 764. [CrossRef]

54. Kravchenko, N.V.; Abramov, V.N.; Semenenko, N.M. Condensation of formic-acid amides and their vinyl analogs with 2, 4, 7-trinitrofluorenes and 2, 4, 5, 7-tetranitrofluorenes. *Russ. J. Org. Chem.* **1989**, *25*, 1938–1945. (in Russian).

55. Mysyk, D.D.; Perepichka, I.F.; Perepichka, D.F.; Bryce, M.R.; Popov, A.F.; Goldenberg, L.M.; Moore, A.J. Electron Acceptors of the Fluorene Series. 9.1 Derivatives of 9-(1,2-Dithiol-3-ylidene)-, 9-(1,3-Dithiol-2-ylidene)-, and 9-(1,3-Selenathiol-2-ylidene)fluorenes: Synthesis, Intramolecular Charge Transfer, and Redox Properties. *J. Org. Chem.* **1999**, *64*, 6937–6950. [CrossRef]

56. Koppel, I.A.; Palm, V.A. *Advances in Linear Free Energy Relationships*; Chapman, N.B., Shorter, J., Eds.; Plenum Press: London, UK, 1972; Chapter 5; p. 203.

57. Palm, V.A. *Fundamentals of Quantitative Theory of Organic Reactions*; Khimiya: Leningrad, Russia, 1977; p. 109. (in Russian)

58. Reichardt, C. *Solvents and Solvent Effects in Organic Chemistry*, 2nd ed.; Wiley-VCH: Weinheim, UK, 1990; pp. 363–371.

59. Ray, A. Effects of temperature and solvent polarity on the interionic charge-transfer interactions in alkylpyridinium bromide. *J. Am. Chem. Soc.* **1971**, *93*, 7146–7149. [CrossRef]

60. Kiprianov, A.I.; Timoshenko, E.S. Contribution to the problem of influence of the solvent on the color of organic pigments 2. *Russ. J. Org. Chem.* **1947**, *17*, 1468–1476. (in Russian).

61. Perepichka, I.F.; Perepichka, D.F.; Bryce, M.R.; Goldenberg, L.M.; Kuzmina, L.G.; Popov, A.F.; Chesney, A.; Moore, A.J.; Howard, J.A.K.; Sokolov, N.I. Fluorene acceptors with intramolecular charge-transfer from 1,3-dithiole donor moieties: Novel electron transport materials. *Chem. Commun.* **1998**, 819–820. [CrossRef]

62. Poelma, J.; Rolland, J. Rethinking digital manufacturing with polymers. *Science* **2017**, *358*, 1384–1385. [CrossRef] [PubMed]

63. Perepichka, D.F.; Perepichka, I.F.; Bryce, M.R.; Sokolov, N.I.; Moore, A.J. π-Extended nitrofluorene-1,3-dithiole chromophore: Enhancing the photoresponse of holographic materials through the balance of intramolecular charge transfer and electron affinity. *J. Mater. Chem.* **2001**, *11*, 1772–1774. [CrossRef]

64. Perepichka, D.F.; Perepichka, I.F.; Bryce, M.R.; Moore, A.J.; Sokolov, N.I. Push-pull dithiole-fluorene acceptors as electron transport materials for holography. *Synth. Met.* **2001**, *121*, 1487–1488. [CrossRef]

65. Perepichka, D.F.; Perepichka, I.F.; Ivasenko, O.; Moore, A.J.; Bryce, M.R.; Kuzmina, L.G.; Batsanov, A.S.; Sokolov, N.I. Combining High Electron Affinity and Intramolecular Charge Transfer in 1,3-Dithiole–Nitrofluorene Push–Pull Diads. *Chem. Eur. J.* **2008**, *14*, 2757–2770. [CrossRef] [PubMed]

66. Bendikov, M.; Perepichka, D.F.; Wudl, F. Tetrathiafulvalenes, Oligoacenenes, and Their Buckminsterfullerene Derivatives: The Brick and Mortar of Organic Electronics. *Chem. Rev.* **2004**, *104*, 4891–4945. [CrossRef] [PubMed]

67. Sanchez, L.; Herranz, M.A.; Martin, N. C60-based dumbbells: Connecting C60 cages through electroactive bridges. *J. Mater. Chem.* **2005**, *15*, 1409–1421. [CrossRef]

68. Echegoyen, L.; Echegoyen, L.E. Electrochemistry of Fullerenes and Their Derivatives. *Acc. Chem. Res.* **1998**, *31*, 593–631. [CrossRef]

69. Adamiak, W.; Opallo, M. Electrochemical redox processes of fullerene C60 and decamethylferrocene dissolved in cast 1,2-dichlorobenzene film in contact with aqueous electrolyte. *J. Electroanal. Chem.* **2010**, *643*, 82–88. [CrossRef]

70. El-Shahawi, M.S.; Omirah, T.A.; Asiri, A.M.; Alsibaai, A.A.; Bashammakh, A.S.; Al-Najjar, H.R. Redox behavior, chromatographic and spectroscopic characterization of some reactive p-conjugated 40-tricyanovinylhydrazone dyes. *RSC Adv.* **2016**, *6*, 5296–5304. [CrossRef]

71. Leclerc, M. Optical and Electrochemical Transducers Based on Functionalized Conjugated Polymers. *Adv. Mater.* **1999**, *11*, 1491–1498. [CrossRef]

72. Garreau, S.; Leclerc, M.; Errien, N.; Louarn, G. Planar-to-Nonplanar Conformational Transition in Thermochromic Polythiophenes: A Spectroscopic Study. *Macromolecules* **2003**, *36*, 692–697. [CrossRef]

73. Borsenberger, P.M.; Weiss, D.S. *Organic Photoreceptors for Xerography*, 1st ed.; Marcel Dekker: New York, NY, USA, 1998; pp. 599–671.

74. Xu, J.; Wen, L.; Zhou, W.; Lv, J.; Guo, Y.; Zhu, M.; Liu, H.; Li, Y.; Jiang, L. Asymmetric and Symmetric Dipole−Dipole Interactions Drive Distinct Aggregation and Emission Behavior of Intramolecular Charge-Transfer Molecules. *J. Phys. Chem. C* **2009**, *113*, 5924–5932. [CrossRef]

75. Turkoglu, G.; Cinar, M.E.; Ozturk, T. Thiophene-Based Organic Semiconductors. *Top. Curr. Chem.* **2017**, *375*, 84. [CrossRef] [PubMed]

76. Moylan, C.R.; McNelis, B.J.; Nathan, L.C.; Marques, M.A.; Hermstad, E.L.; Brichler, B.A. Challenging the Auxiliary Donor Effect on Molecular Hyperpolarizability in Thiophene-Containing Nonlinear Chromophores: X-ray Crystallographic and Optical Measurements on Two New Isomeric Chromophores. *J. Org. Chem.* **2004**, *69*, 8239–8243. [CrossRef] [PubMed]

77. Skabara, P.J.; Serebryakov, I.M.; Perepichka, I.F. Electron acceptors of the fluorene series. Part 8.1 Electrochemical and intramolecular charge transfer studies of thiophene functionalised fluorenes. *J. Chem. Soc. Perkin Trans.* **1999**, *2*, 505–513. [CrossRef]

78. Skabara, P.J.; Müllen, K.; Bryce, M.R.; Howard, J.A.K.; Batsanov, A.S. The synthesis, redox properties and X-ray crystal structures of two new tetrathiafulvalene-thiophene donors. *J. Mater. Chem.* **1998**, *8*, 1719–1724. [CrossRef]

79. Charlton, A.; Underhill, A.E.; Williams, G.; Kalaji, M.; Murphy, P.J.; Abdul Malik, K.M.; Hursthouse, M.B. Thiophene-Functionalized TTF π-Electron Donors as Potential Precursors to Conducting Polymers and Organic Metals: Synthesis, Properties, Structure, and Electropolymerization Studies. *J. Org. Chem.* **1997**, *62*, 3098–3102. [CrossRef] [PubMed]

80. Skabara, P.J.; Serebryakova, I.M.; Perepichka, I.F. Synthesis and electropolymerisation of thiophene functionalised fluorenes. *Synth. Met.* **1999**, *102*, 1336–1337. [CrossRef]

81. Skabara, P.J.; Serebryakov, I.M.; Perepichka, I.F.; Serdar Sariciftci, N.; Neugebauer, H.; Cravino, A. Toward Controlled Donor−Acceptor Interactions in Noncomposite Polymeric Materials: Synthesis and Characterization of a Novel Polythiophene Incorporating π-Conjugated 1,3-Dithiole-2-ylidenefluorene Units as Strong D−A Components. *Macromolecules* **2001**, *34*, 2232–2241. [CrossRef]

82. Perepichka, I.F.; Popov, A.F.; Orekhova, T.V.; Bryce, M.R.; Vdovichenko, A.N.; Batsanov, A.S.; Goldenberg, L.M.; Howard, J.A.K.; Sokolov, N.I.; Megson, J.L. Electron acceptors of the fluorene series. Part 5. Intramolecular charge transfer in nitro-substituted 9-(aminomethylene)fluorenes. *J. Chem. Soc. Perkin Trans.* **1996**, *2*, 2453–2469. [CrossRef]

83. Torrance, J.B. An Overview of Organic Charge-Transfer Solids: Insulators, Metals, and the Neutral-Ionic Transition. *Mol. Cryst. Liq. Cryst.* **1985**, *126*, 55–67. [CrossRef]

84. Hertler, W.R.; Hartzler, H.D.; Acker, D.S.; Benson, R.E. Substituted Quinodimethans. III. Displacement Reactions of 7,7,8,8-Tetracyanoquinodimethan. *J. Am. Chem. Soc.* **1962**, *84*, 3387–3393. [CrossRef]

85. McKusick, B.C.; Heckert, R.E.; Cairus, T.L.; Coffman, D.D.; Mower, H.F. Cyanocarbon Chemistry. VI.1 Tricyanovinylamines. *J. Am. Chem. Soc.* **1958**, *80*, 2806–2815. [CrossRef]

86. Perepichka, I.F.; Popov, A.F.; Artyomova, T.V.; Vdovichenko, A.N.; Bryce, M.R.; Batsanov, A.S.; Howard, J.A.K.; Megson, J.L. Fluorene derivatives with intramolecular charge-transfer: Exceptionally easy rotation around the double C(9)[double bond, length half m-dash]C(α) bond in nitro-substituted 9-aminomethylenefluorenes. *J. Chem. Soc. Perkin Trans.* **1995**, *2*, 3–5. [CrossRef]

87. Aviram, A.; Ratner, M. Molecular rectifiers. *Chem. Phys. Lett.* **1974**, *29*, 277–283. [CrossRef]

88. Panetta, C.A.; Heimer, N.E.; Hussey, C.L.; Metzger, R.M. Functionalized Tetracyanoquinodimethane-type Electron Acceptors: Suitable Precursors for D-σ-A Materials. *Synlett* **1991**, 301–309. [CrossRef]

89. Bryce, M.R. Tetrathiafulvalenes as π-Electron Donors for Intramolecular Charge-Transfer Materials. *Adv. Mater.* **1999**, *11*, 11–23. [CrossRef]

90. Bryce, M.R. Functionalised tetrathiafulvalenes: New applications as versatile π-electron systems in materials chemistry. *J. Mater. Chem.* **2000**, *10*, 589–598. [CrossRef]

91. Perepichka, D.F.; Bryce, M.R.; Pearson, C.; Petty, M.C.; McInnes, E.J.L.; Zhao, J.P. A Covalent Tetrathiafulvalene–Tetracyanoquinodimethane Diad: Extremely Low HOMO–LUMO Gap, Thermoexcited Electron Transfer, and High-Quality Langmuir–Blodgett Films. *Angew. Chem. Int. Ed.* **2003**, *42*, 4636–4639. [CrossRef] [PubMed]

92. Tsiperman, E.; Becker, J.Y.; Khodorkovsky, V.; Shames, A.; Shapiro, L. A Rigid Neutral Molecule Involving TTF and TCNQ Moieties with Intrinsic Charge- and Electron-Transfer Properties that Depend on the Polarity of the Solvent. *Angew. Chem. Int. Ed.* **2005**, *44*, 4015–4018. [CrossRef] [PubMed]

93. Guégano, X.; Kanibolotsky, A.L.; Blum, C.; Mertens, S.F.L.; Liu, S.-X.; Neels, A.; Hagemann, H.; Skabara, P.J.; Leutwyler, S.; Wandlowski, T.; Hauser, A.; Decurtins, S. Pronounced Electrochemical Amphotericity of a Fused Donor–Acceptor Compound: A Planar Merge of TTF with a TCNQ-Type Bithienoquinoxaline. *Chem. Eur. J.* **2009**, *15*, 63–66. [CrossRef] [PubMed]

94. Dumur, F.; Guégano, X.; Gautier, N.; Liu, S.-X.; Neels, A.; Decurtins, S.; Hudhomme, P. Approaches to Fused Tetrathiafulvalene/Tetracyanoquinodimethane Systems. *Eur. J. Org. Chem.* **2009**, 6341–6354. [CrossRef]

95. Tsiperman, E.; Regev, L.; Becker, J.Y.; Bernstein, J.; Ellern, A.; Khodorkovsky, V.; Shames, A.; Shapiro, A.L. Synthesis of a novel rigid tetrathiafulvalene-σ-p-benzoquinone diad (TTF-σ-Q) with inherent structural configuration suitable for intramolecular charge-transfer. *Chem. Commun.* **1999**, 1125–1126. [CrossRef]

96. Sheib, S.; Cava, M.P.; Baldwin, J.W.; Metzger, R.M. In Search of Molecular Rectifiers. The Donor$-\sigma-$Acceptor System Derived from Triptycenequinone and Tetrathiafulvalene. *J. Org. Chem.* **1998**, *63*, 1198–1204. [CrossRef]

97. Martin, N.; Sanchez, L.; Herranz, M.A.; Guldi, D.M. Evidence for Two Separate One-Electron Transfer Events in Excited Fulleropyrrolidine Dyads Containing Tetrathiafulvalene (TTF). *J. Phys. Chem. A* **2000**, *104*, 4648–4657. [CrossRef]

98. Perepichka, D.F.; Bryce, M.R.; Batsanov, A.S.; McInnes, E.J.L.; Zhao, J.P.; Farley, R.D. Engineering a Remarkably Low HOMO–LUMO Gap by Covalent Linkage of a Strong π-Donor and a π-Acceptor—Tetrathiafulvalene-σ-Polynitrofluorene Diads: Their Amphoteric Redox Behavior, Electron Transfer and Spectroscopic Properties. *Chem. Eur. J.* **2002**, *8*, 4656–4669. [CrossRef]

99. Boyd, R.H.; Phillips, W.D. Solution dimerization of the tetracyanoquinodimethane ion radical. *J. Phys. Chem.* **1965**, *43*, 2927–2929. [CrossRef]

100. Perepichka, I.F.; Perepichka, D.F.; Bryce, M.R.; Chesne, A.; Popov, A.F.; Khodorkovsky, V.; Meshulamd, G.; Kotler, Z. Push-pull fluorene acceptors with ferrocene donor moiety. *Synth. Met.* **1999**, *102*, 1558–1559. [CrossRef]

101. Perepichka, D.F.; Perepichka, I.F.; Popov, A.F.; Bryce, M.R.; Batsanov, A.S.; Chesney, A.; Howard, J.A.K.; Sokolov, N.I. Electron acceptors of the fluorene series. Part 12. 9-(Metalloceneylidene)nitrofluorene derivatives of Fc–NF, NF–Fc–NF, and NF–Rc–NF types, and the vinylogues Fc—NF: Synthesis, characterisation, intramolecular charge transfer, redox properties and X-ray structures for three fluorene–ferrocene derivatives. *J. Organomet. Chem.* **2001**, *637*, 445–462.

Article

Low Temperature Solution-Processable 3D-Patterned Charge Recombination Layer for Organic Tandem Solar Cells

Jin Woo Choi [1], Jong Woo Jin [1], Denis Tondelier [1], Yvan Bonnassieux [1] and Bernard Geffroy [1,2,*]

[1] LPICM, CNRS, Ecole Polytechnique, Université Paris Saclay, 91128 Palaiseau, France; jinwoo.choi@polytechnique.edu (J.W.C.); jongwoo.jin@polytechnique.edu (J.W.J.); denis.tondelier@polytechnique.edu (D.T.); yvan.bonnassieux@polytechnique.edu (Y.B.)

[2] LICSEN, NIMBE, CEA, CNRS, Université Paris-Saclay, CEA Saclay, 91191 Gif-sur-Yvette Cedex, France

* Correspondence: bernard.geffroy@polytechnique.edu

Received: 7 December 2018; Accepted: 25 December 2018; Published: 7 January 2019

Abstract: We propose a novel method to pattern the charge recombination layer (CRL) with a low-temperature solution-processable ZnO layer (under 150 °C) for organic solar cell applications. Due to the optimal drying process and thermal annealing condition, ZnO sol-gel particles formed a three-Dimensional (3D) structure without using a high temperature or ramping method. The generated 3D nano-ripple pattern showed a height of around 120 nm, and a valley-to-valley distance of about 500 nm. Based on this newly developed ZnO nano-ripple patterning technique, it was possible to pattern the CRL without damaging the underneath layers in tandem structure. The use of nano-ripple patterned ZnO as the part of CRL, led to the concomitant improvement of the power conversion efficiency (PCE) of about 30%, compared with non-patterned CRL device.

Keywords: organic tandem solar cell; 3D nano-ripple pattern; ZnO sol-gel; charge recombination layer; low temperature solution process

---

1. Introduction

Finding alternatives for the current energy sources (i.e., burning fossil fuels, nuclear materials) has become one of the most important societal challenges for relieving the environmental pollution problem [1]. One of the most promising next-generation energy sources is solar energy, which can be converted to electric power via photovoltaic technology. Currently industrialized photovoltaic panels are based on inorganic materials such as silicon [2–4]. Recently, organic solar cells (OSCs) emerged as an alternative to inorganic photovoltaics devices [5–7]. The merits of OSC technology are: a low-cost solution process, a low temperature process, flexibility, and a tailorable material for further improvement. Recently, the champion single-junction OSC has reached a power conversion efficiency (PCE) of 12.6% [8]. However, it is still necessary for improving PCE and air-stability for large-scale commercialization.

One of the reasons for the low PCE is the narrow light absorption range of organic materials. A tandem solar cell structure, where two or more single-junction cells with complementary absorption spectra are connected in tandem, can be a promising design to overcome the limitations of single cells [9–11]. A tandem structure offers several advantages: (1) a broad absorption spectrum due to the usage of complementary absorbing materials; (2) summation of the open circuit voltage (V_{OC}) of each sub-cell; (3) a reasonable fill factor (FF) due to higher optical density over a wider fraction of solar spectrum than that of single cell without increasing internal resistance. To maximize these advantages, the tandem device requires the qualified charge recombination layer (CRL) to simultaneously act as the anode for one of two adjacent sub-cells, and as the cathode for the other [12]. The CRL should have low electrical resistance, high optical transparency in the visible range, and a low barrier for charge recombination. Furthermore, the layer should be able to protect the lower layers during the remaining solution fabrication process.

In 2007, the CRL made with a metal oxide material was suggested [13]. The advantage of using a metal oxide as a part of CRL is to minimize the absorption at visible wavelengths. The classical structure of CRL is based on 0.5 nm LiF/1 nm Al/3 nm WoO$_3$, used as multilayers. The first all-solution-processable tandem OSC was reported by Kim et al., where poly(3,4-ethylenedioxythiophene) polystyrene sulfonate (PEDOT:PSS) and TiO$_X$ were used to form CRL [14]. In 2013, the first inverted tandem OSC that broke 10% efficiency was reported by You et al., with PEDOT:PSS and ZnO being used as CRL [15]. As shown above, new efficient CRLs with a large work function difference between two opposite interfaces have been developed continuously. Therefore, is high V$_{OC}$ of the tandem device without loss-in-sum of the V$_{OC}$ of the component cells is achieved. However, maximizing the short circuit current of tandem device is still under development, due to the difficulties in the current matching between the component cells. The current of tandem device is limited by the cell that produces the lower current. Therefore, it is important to increase the short circuit current of each component cell by enhancing the charge extraction properties of CRL. However, to our knowledge, there has been no reported research on the tandem device with patterned CRL, even though the 3D pattern could enhance the charge extraction capability of the CRL.

In the single junction solar cell, patterning of the charge-collecting layer (CCL) has been developed widely to maximize the charge extraction ratio [16–21]. However, previously proposed fabrication techniques for growing nano-wires, nano-rods, and nanoporous layers are not compatible with tandem structure fabrication. High thermal annealing conditions or vacuum processes should be avoided for all solution-processable tandem devices, to prevent the degradation of the films underneath. Meanwhile, a solution-processable ZnO nano-ripple pattern was firstly shown in a single sub-device by Yang et al. in 2009, by using a ramping thermal annealing method [22]. By using patterned ZnO CCL, the PCE improved by about 25% compared with device-containing planar CCL. However, the high thermal annealing treatment condition of this technique (275–350 °C) limited the use of this ripple patterning only in single inverted OSCs [22–24].

Herein, we found the optimal process conditions for nano-ripple patterning of the ZnO film in low-temperature conditions. After confirming the characteristics of ZnO in the single sub-cell device, the layer was introduced as a part of CRL in the tandem device. A newly developed low-temperature process technique actualized the patterning on the CRL without damaging the films underneath. As compared to the non-patterned device, the tandem device with ZnO ripple-patterned CRL showed a PCE improvement of about 30%.

2. Experimental Section

As shown in Figure 1, for electron-only devices, ZnO films with different pattern sizes, single-junction OSCs, and tandem OSCs with a ZnO layer were fabricated and analyzed. The detailed fabrication methods are described below.

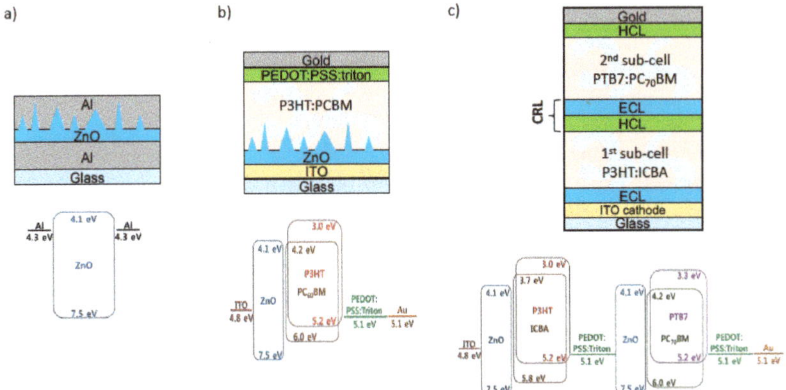

Figure 1. Device structures used in this study, with relevant energy levels of the different layers. (**a**) Electron only device, (**b**) single solar cell with a nano-ripple pattern, (**c**) tandem solar cell. ECL (electron collecting layer) and HCL (hole-collecting layer) are the ZnO nano-ripple layer and the PEDOT:PSS:Triton layer respectively.

2.1. Preparation of Solutions

A ZnO$_X$ sol-gel solution was prepared by the following method. Zinc acetate dehydrate (ZnO precursor) at 1.5 M concentration was prepared in 2-methoxyethanol, followed by adding 0.1 mL of ethanolamine. Two types of PEDOT:PSS (Al4083 and PH1000, respectively) with surfactant were used as the hole-collecting layer. A volume of 100 µL of surfactant Triton was mixed with 4 mL of 1:1 volume ratio of AI4083 and PH1000. The ratio of PEDOT:PSS was 1:6 for Al4083, and 1:2.5 for PH1000. PH1000 contains higher amount of PEDOT, and thus has a higher conductivity than AI4083. AI4083 and PH1000 were purchased from Bayer (Leverkusen, Germany). In solar cells, poly(3-hexylthiophene) (P3HT) and poly[[4,8-bis[(2-ethylhexyl)oxy]benzo[1,2-b:4,5-b']dithiophene-2,6-diyl][3-fluoro-2-[(2-ethylhexyl)carbonyl]thieno[3,4-b]thiophenediyl]] (PTB7) were used as the donor materials, while 1′,1″,4′,4″-tetrahydro-di[1,4]methanonaphthaleno[1,2:2′,3′,56,60:2″,3″][5,6]fullerene-C60 (ICBA), 6,6-phenyl C61-butyric acid methyl ester (PCBM), and 6,6-phenyl C71-butyric acid methyl ester (PC$_{70}$BM) were used as acceptor materials. P3HT, PTB7, and fullerene derivatives were purchased from Ossila (Sheffield, UK). For a single solar cell (Figure 1b), a blend of P3HT and PCBM (1:1 weight ratio) was used as the active layer by dissolving 40 mg of P3HT:PCBM in 1 mL 1,2-dichlorobenzene (DCB, Sigma Aldrich, St. Louis, MO, USA). For the tandem solar cell (Figure 1c), the first sub-cell (bottom cell, Figure 1c) was based on a blend of P3HT and ICBA, and the second sub-cell (top cell, Figure 1c) was a blend of PTB7 and PC$_{70}$BM. The P3HT:ICBA blend solution was prepared by dissolving 40 mg of P3HT:ICBA (1:1 weight ratio) in 1 mL of DCB. For second sub-cell active layer, the solution was prepared by dissolving 1:1.5 weight ratio of PTB7:PC$_{70}$BM in DCB with a total concentration of 25 mg/mL. All prepared solutions were stirred for 12 h in a N$_2$ glove box.

2.2. Electron-Only Device

An aluminum layer of 100 nm thickness was deposited onto glass substrates by thermal evaporation. The deposition rate of Al was 0.25 Å/s under 5×10^{-7} Torr process pressure. Prepared ZnO sol-gel solution was spin-coated on the Al deposited glass through a 0.45 µm pore polyvinylidene fluoride (PVDF) filter at 1000 rpm for 60 s. The obtained films were dried in ambient air over 5, 10, 30, 60, and 180 min, respectively, and then thermally annealed over 10 min at 125, 150, 175, and 200 °C, respectively. The final thickness of the films was 400 nm. For top electrode, 100 nm thick aluminum layer was thermally evaporated through a shadow mask, defining a device area of 0.28 cm^2. The deposition condition of Al was same as the bottom electrode. ZnO film obtained by ramping method, as described by Yang group, was also prepared for comparison [21]. The current-voltage (I-V) characteristics of the ZnO electron-only device were measured by a Keithley 4200 (Tektronix, Beaverton, OR, USA). Atomic force microscopy (AFM) was used for the imaging of the 3D pattern of the fabricated samples.

2.3. Organic Solar Cell Fabrication and Characterization

OSCs were fabricated on ITO substrates (sheet resistance 20 Ω/square) purchased from Xinyan Technologies (Kowloon, Hong Kong). Pre-patterned ITO coated glass substrates were cleaned in an ultrasonic cleaner with deionized water and isopropyl alcohol (IPA) for 20 min each.

For single solar cell (Figure 1b), prepared ZnO sol-gel solution was deposited onto the ITO layer by spin coating (3000 rpm for 60 s) through a 0.45 μm pore PVDF filter in a N_2 glove box. The substrates were directly transferred outside the glove box and dried in ambient air during different times (0, 5, 10, 30, 60, and 180 min, respectively) and then thermally annealed at 125 °C over 10 min, in order to form nano-ripple structures. The final thickness of the fabricated film was 200 nm. The samples were then transferred into N_2 glove box to deposit the following layers. A P3HT:PCBM solution was filtered through a 0.45 μm pore polytetrafluoroethylene (PTFE) filter, and spin-coated in two steps (500 rpm for 30 s, and then 1000 rpm for 45 s). The films were thermally annealed at 145 °C for 10 min. The thickness of this layer was approximately 100 nm after annealing. Then, PEDOT:PSS:Triton solution was deposited at 1000 rpm for 60 s. The thickness of this layer was 120 nm after annealing at 120 °C for 2 min.

For tandem cells (Figure 1c), the ZnO nano-ripple layer (ECL), the first sub-cell active layer, and PEDOT:PSS:Triton layer (HCL) were fabricated in the same way as reported above. PTB7:PC$_{70}$BM was used as active layer in the second sub-cell. The PTB7:PC$_{70}$BM solution was filtered through a 0.45 μm pore PTFE filter, and spin-coated at 700 rpm for 60 s. The layer was annealed in DCB solvent vapor for 5 min. The thickness of this layer was approximately 80 nm. Finally, a gold electrode (anode) was thermally evaporated through a shadow mask, defining a 0.28 cm^2 active surface area. The deposition rate of Au was 0.5 Å/s under 5×10^{-6} Torr process pressure.

The current density-voltage (J-V) characteristics of the OSCs were measured in a N_2 glove box using a source meter (Keithley 2635, Tektronix, Beaverton, OR, USA) in the dark and under illumination. An Air-Mass 1.5 (AM 1.5) solar simulator with 100 mW cm^{-2} was used as the light source.

3. Results and Discussion

3.1. ZnO Electron Collection Layer

To find the optimal annealing temperature for ZnO films, electron-only devices were fabricated to evaluate the conductivity of ZnO films under different thermal annealing conditions [25]. Figure 2 shows the I-V characteristics of Al/ZnO/Al electron-only device, with 400 nm thick ZnO. Table 1 summarizes the conductivity extracted from the I-V characteristics. The film made with the ramping fabrication method shows the highest conductivity among all films, but the difference is weak, and the conductivity values are in the same order of magnitude. A factor of 3 was observed between the lowest temperature annealing temperature (125 °C) and the ramping method (350 °C). It is obvious that the conductance of the ZnO thin film becomes higher when the film was annealed at higher temperature. Generally, a high annealing temperature provides a better crystallinity of the films, which directly impacts on the conductivity [26,27]. The reaction phenomena of ZnO sol-gel is reported in a previous report [28]. The ZnO precursor (zinc acetate dehydrate) film is highly resistive, because of the acetate functional group. Therefore, an annealing process is required to induce the reaction between the ZnO precursor with 2-methoxyethanol and the oxidation of a ZnO precursor in air. However, thermal annealing treatments of higher than 200 °C were not considered in this study, since the high annealing temperature of CRL could cause damage on the underlying layers in a tandem structure. It is then shown that the annealing temperature could be lowered to 125 °C.

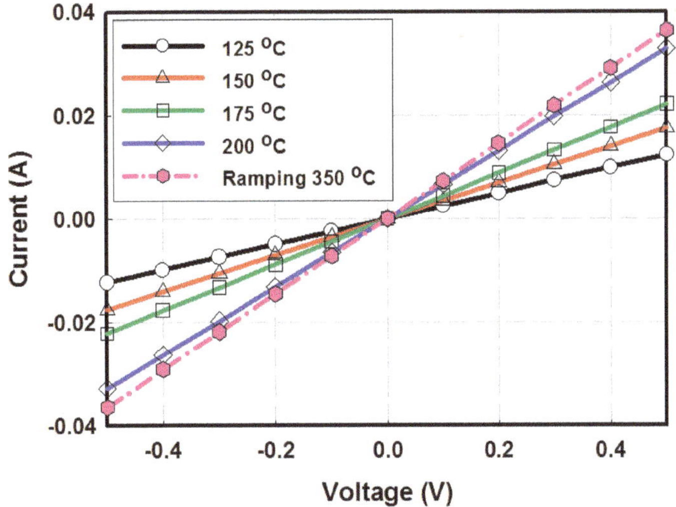

Figure 2. I-V characteristics of an Al (100 nm)/ZnO (400 nm)/Al (100 nm) electron-only diode device with a ZnO thickness of 400 nm.

Table 1. Conductivity value of the ZnO films under different annealing temperatures.

Temperature	125 °C	150 °C	175 °C	200 °C	Ramping 350 °C
Conductivity (S/m)	2.5×10^{-3}	3.5×10^{-3}	4.4×10^{-3}	6.7×10^{-3}	7.3×10^{-3}

3.2. ZnO Ripple Pattern

Based on the idea that ZnO ripple pattern growth mechanism is analogous to the coffee ring effect [27,29,30] and the solvent annealing technique [31], ZnO films were dried in the air at ambient temperature over various time periods (from 0 to 3 h) before annealing. After that, they were placed on a hot plate for thermal treatment over 10 minutes at 125, 150, 175, and 200 °C, respectively. Peak-to-valley roughness values (R_{PtoV}) of patterned ZnO films, measured by AFM, are summarized in Table 2. The highest R_{PtoV} values for each annealing temperature were as follows: 110 nm at 125 °C, 122 nm at 150 °C, and 113 nm at 175 °C, achieved when the drying time was 60, 10, and 5 min, respectively. Devices with a short drying time require a high annealing temperature for the optimum pattern, i.e., the highest R_{PtoV} value. This result is reasonable, because short dried samples has more remaining solvent [32]. The sample will need a higher temperature to remove the remaining solvent, and consequently, to form the pattern. During the drying process, the films will partially dry, and small ZnO particles (cross-linked ZnO precursors) begin to form inside the film. These particles will travel, following the convection current of the solvent during thermal annealing of the film. Changing the drying time affects the initial size of seed crystal particles and the amount of the remaining solvent, which modifies the convection flow and the coffee ring effect [29,32]. The device with the highest R_{PtoV} was achieved at 150 °C after 10 min of drying time. The R_{PtoV} values were abruptly decreased with a long drying time (3 h) and a high annealing temperature (200 °C). The lack of remaining solvent due to the long drying process or fast evaporation at high temperature hinders the convection flow, and consequently, the rearrangement of ZnO precursor to form a 3D pattern.

The AFM images of the best nano-ripple pattern fabricated by a low temperature process, and the pattern obtained by ramping method, are shown in Figure 3. Pattern sizes with the novel low temperature method (Figure 3a) showed similar R_{PtoV} value to that achieved by the ramping method (Figure 3b). R_{PtoV} value of ZnO film fabricated by ramping method was 135 and 120 nm for drying time of 10 min followed by annealing temperature of 150 °C during 10 min. Finally, we succeed on the fabrication of 3D nano-patterned ZnO film, without using a high temperature, by introducing the drying step in the process.

Table 2. R_{PtoV} value of ZnO films with different drying times and annealing temperatures (units in nm).

Drying Time	Annealing Temperature			
	125 °C	**150 °C**	**175 °C**	**200 °C**
0 min	78	109	93	21
5 min	84	111	113	22
10 min	96	122	110	19
30 min	102	106	99	21
60 min	110	98	93	22
3 h	43	87	83	21

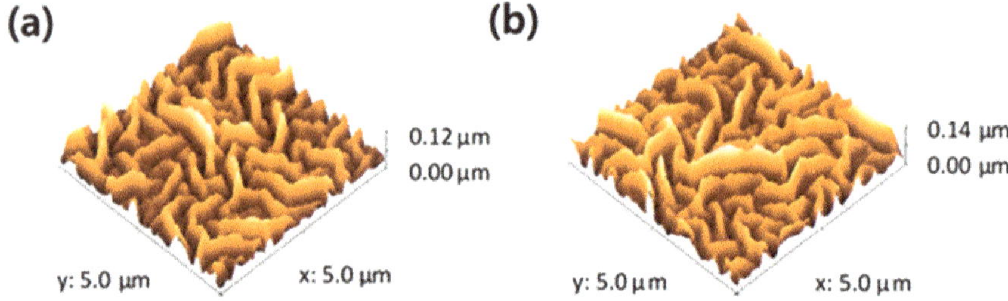

Figure 3. Atomic force microscopy (AFM) images of the ZnO nano-ripple, (**a**) the low temperature processed (10 minutes of drying time and annealing temperature of 150 °C) and (**b**) the ramping method.

3.3. Single Junction Solar Cell with A 3D Nano-Patterned ZnO Layer

The inverted devices have a structure of ITO/ZnO/P3HT:PCBM (200 nm)/PEDOT:PSS (100 nm)/Al (100 nm) with a different ripple size of the ZnO layer. Figure 4 shows J-V characteristics of the inverted devices made with P3HT:PCBM under 100 mW/cm^2 illumination. The performance characteristics of the OSCs are summarized in Table 3. Devices A to D were fabricated in a low-temperature process, with an annealing temperature of 150 °C for 10 min and different drying times. The drying time was 3000, 300, 60, and 10 min for devices A, B, C, and D, respectively. As previously shown in Table 2, the drying time modifies the pattern size (R_{PtoV}) for a fixed temperature. Only drying time was controlled to change the 3D pattern size (R_{PtoV}), in order to exclude the effects of the annealing temperature on the film properties, such as conductivity. The J-V curve that is labeled device E shows the characteristics of OSC, with the ZnO layer being made by the ramping method. Even though the ramping method could give a slightly better conductivity as shown in Table 1, the overall characteristic of device D shows similar characteristic with device E. The short circuit current (J_{SC}) and FF increase as the ZnO pattern size increases, while V_{OC} does not show a significant change. As a result, the device with 120 nm R_{PtoV} has the best performance, with 3.4% of PCE. The improvements on the J_{SC} and FF are based on the efficient charge collection properties, by introducing a patterned ZnO layer. Due to the efficient carrier extraction, the space charge-inducing resistance near ECL will be reduced. When the pattern size of ZnO increases from 80 to 120 nm, an expectable result is a decrease in the series resistance. The series resistances extracted by the slope of the J-V curve at the V_{OC} were 27, 25, 21, and 21 Ω·cm^2 for device A, B, C, and D, respectively.

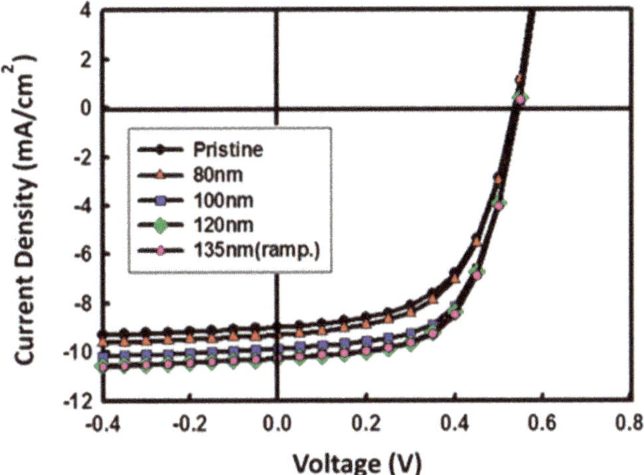

Figure 4. Current–voltage characteristics of the solar cells with different nano-patterned sizes of the ZnO electron-collecting layer (ECL). (ramp.) stands for the ramping method.

Table 3. The summary of the key parameters of the single junction organic solar cells (OSCs), with different nano-ripple pattern sizes. Devices A to D were fabricated by the low-temperature method, while device E was fabricated by the ramping method.

Device X	Pattern Size (nm)	Voc (V)	Jsc (mA/cm^2)	FF (%)	PCE (%)	Rs ($\Omega \cdot$cm^2)
A	0	0.54	−9.0	56	2.7	27
B	80	0.54	−9.3	56	2.8	25
C	100	0.55	−9.9	61	3.3	21
D	120	0.55	−10.3	60	3.4	21
E	135	0.55	−10.3	61	3.4	21

To assess the effect of ZnO layer on the electrical properties of the device more deeply, we examined the charge collection probability ($P_{collection}$) in the cells, according to the method proposed by Kyaw et al. [33]. A photo-generated current (J_{ph}) is saturated at the high internal voltage (V_{int}) region, due to the large enough internal field extracted all generated charges. Therefore, the saturated photocurrent ($J_{ph,saturation}$) can be written as following equation, while the value is limited only by the number of absorbed photons:

$$J_{ph,saturation} = qLG_{max} \tag{1}$$

q is the elementary charge, L is the thickness of the active layer, and G_{max} is the maximum photoinduced carrier generation rate per unit volume. However, in a low internal voltage region, the J_{ph} is proportional to the $P_{collection}$ in the cells. Therefore, the equation 1 can be written as:

$$J_{ph,saturation} = qLG_{max} P_{collection} \tag{2}$$

From Equations (1) and (2), $P_{collection}$ can be calculated by normalizing J_{ph} with $J_{ph,saturation}$.

Figure 5 shows the charge collection probability with respect to internal voltage (V_{int}) under illumination of 100 mW/cm^2. As shown in Figure 5, the overall charge collection probability of the inverted devices with ZnO increased, as the pattern size increases. The increment in charge collection probability is more significant at low V_{int} (high applied voltage). The observed increment of charge collection probability in the nano-patterned ZnO devices is well-matched with the J_{SC} characteristic of the OSCs.

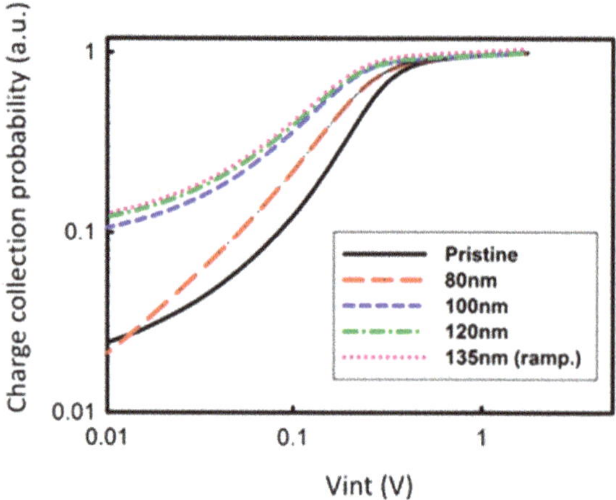

Figure 5. Charge collection probability as a function of the internal voltage for cells with different ZnO nano-pattern size.

3.4. Tandem Solar Cell with Low-Temperature Processed ZnO

Photoactive layers with complementary absorption range were used as sub-cells of the tandem device. P3HT:ICBA was selected as the active layer for the bottom cell (device F and G). ICBA was selected as the acceptor, because the P3HT:ICBA active layer forms its optimal morphology in higher annealing temperature when compared with P3HT:PCBM [34–36]. Therefore, the P3HT:ICBA system promises a lower thermal degradation for tandem device fabrication, compared with the P3HT:PCBM system. The PTB7:PC$_{70}$BM active layer is selected for the top cell (device H and I). Instead of thermal annealing, the solvent annealing method was used for the PTB7:PC$_{70}$BM active layer, to minimize the thermal effect on the underneath layers in the tandem device process [37]. The sub-cells have the structure: ITO/ZnO (150 nm)/P3HT:ICBA (100 nm) or PTB7:PC$_{70}$BM (150 nm)/PEDOT:PSS (120 nm)/Au (100 nm). The details of structure and photovoltaic devices performance are summarized in Table 4. Figure 6 presents the effect of the ZnO ripple pattern on the performance of each sub-cell. Among the devices F to I, devices with patterned ZnO show better J$_{SC}$ and PCE. Likewise in the device with P3HT:PCBM, the increment of charge collection probability in the nano-patterned ZnO layer is the origin of these improvements. The inverted tandem devices have a structure of ITO/ZnO (150 nm)/P3HT:ICBA (100 nm)/PEDOT:PSS (120 nm)/ZnO (150 nm)/PTB7:PC$_{70}$BM (150 nm)/PEDOT:PSS/Au (100 nm). The V$_{OC}$ values of fabricated tandem devices are much higher than V$_{OC}$ of each sub-cell (Table 4). This means that the sub-cells were series connected through CRL. V$_{OC}$ of P3HT:ICBA device and PTB7:PC$_{70}$BM device is 0.78 and 0.73 V, respectively. Therefore, ideally expected V$_{OC}$ for tandem device is 1.5 V. Comparing this ideal value with V$_{OC}$ of the fabricated tandem devices, there was only slight loss at around 0.1 to 0.2 V. A tandem device with a 3D nano-patterned ZnO (device K) shows a higher V$_{OC}$, J$_{SC}$, and FF than the device with pristine ZnO (device J). A better work function matching of CRL with active layers, efficient electron extraction, and electron-hole balance in the device are the reasons for this improvement [38,39]. A highly efficient carrier recombination characteristic of the CRL in a patterned ZnO layer reduced charge accumulation effect. Finally, the best device achieved a V$_{OC}$ of 1.38 V, a J$_{SC}$ of −7.2 mA/cm^2, a FF of 49%, and PCE of 4.8%.

Table 4. The summary of the key parameters of the OSCs extracted from the J-V measurements. ECL stands for the electron collecting layer (ZnO layer).

Device	Structure	ECL (ZnO Layer)	V_{OC} (V)	Jsc (mA/cm^2)	FF (%)	PCE (%)
F	First sub-cell	Non-patterned	0.79	−7.6	51	3.1
G		Nano-ripple	0.78	−8.8	50	3.5
H	Second sub-cell	Non-patterned	0.73	−10.2	48	3.6
I		Nano-ripple	0.73	−11.9	48	4.2
J	Tandem cell	Non-patterned	1.28	−6.5	44	3.7
K		Nano-ripple	1.38	−7.2	49	4.8

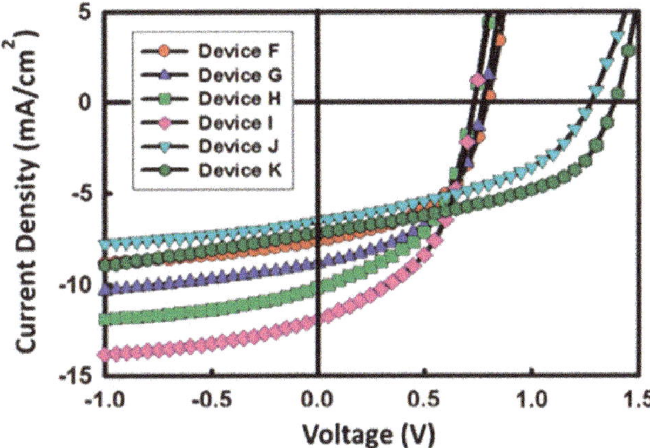

Figure 6. The J-V characteristics of the OSC devices with and without a ZnO patterned layer as ECL.

4. Conclusions

We developed a new method for patterning the ZnO layer by a low temperature (under 150 °C) solution process. A solution-processable ZnO layer shows proper electrical characteristics for OSCs even with a low temperature process. Nano-ripple pattern was fabricated to maximize the interface between the ZnO layer and the photoactive layer. In a single-junction device, PCE was about 30% higher when the ZnO layer was patterned. Finally, developed low-temperature processable nano-patterned ZnO was introduced in the tandem structure. The best tandem device has V_{OC} of 1.38 V and PCE of 4.8%, showing nearby 30% PCE improvement, compared to the non-patterned injection layer. It is shown that main improvement results from the charge extraction probability improvement, due to the nano-ripple patterned layer.

Author Contributions: J.W.C., Y.B., and B.G. conceived and designed the experiments; J.W.C., J.W.J., and D.T. performed the experiments; J.W.C. and B.G. analyzed the data; J.W.C. and B.G. wrote the paper.

Acknowledgments: J.C. would like to thank ED Polytechnique for the PhD funding.

References

1. BP. *BP Statistical Review of World Energy*; British Petroleum Company: London, UK, 2014; pp. 1–48.

2. El Chaar, L.; Lamont, L.A.; El Zein, N. Review of Photovoltaic Technologies. *Renew. Sustain. Energy Rev.* **2011**, *15*, 2165–2175. [CrossRef]

3. Becquerel, E. Mémoire sur les effets électriques produits sous l'influence des rayons solaires. *C. R. Acad. Sci. Paris* **1839**, *9*, 561–567.

4. Chapin, D.M.; Fuller, C.S.; Pearson, G.L. A New Silicon p-n Junction Photocell for Converting Solar Radiation into Electrical Power. *J. Appl. Phys.* **1954**, *25*, 676–677. [CrossRef]

5. Hoppe, H.; Sariciftci, N.S. Organic solar cells: An overview. *J. Mater. Res.* **2011**, *19*, 1924–1945. [CrossRef]

6. Wohrle, D.; Meissner, D. Organic Solar Cells. *Adv. Mater.* **1991**, *3*, 129–138. [CrossRef]

7. Gregg, B.A. Excitonic solar cells. *J. Phys. Chem. B* **2003**, *107*, 4688–4698. [CrossRef]

8. Available online: https://www.nrel.gov/pv/assets/pdfs/pv-efficiencies-07-17-2018.pdf (accessed on 7 December 2018).

9. Cheyns, D.; Rand, B.P.; Heremans, P. Organic tandem solar cells with complementary absorbing layers and a high open-circuit voltage. *Appl. Phys. Lett.* **2010**, *97*, 033301. [CrossRef]

10. Dennler, G.; Prall, H.J.; Koeppe, R.; Egginger, M.; Autengruber, R.; Sariciftci, N.S. Enhanced spectral coverage in tandem organic solar cells. *Appl. Phys. Lett.* **2006**, *89*, 073502. [CrossRef]

11. Dou, L.; You, J.; Yang, J.; Chen, C.C.; He, Y.; Murase, S.; Moriarty Emery, K.; Li, G.; Yang, Y. Tandem polymer solar cells featuring a spectrally matched low-bandgap polymer. *Nat. Photonics* **2012**, *6*, 180–185. [CrossRef]

12. Zhou, Y.; Fuentes-Hernandez, C.; Shim, J.W.; Khan, T.M.; Kippelen, B. High performance polymeric charge recombination layer for organic tandem solar cells. *Energy Environ. Sci.* **2012**, *5*, 9827–9832. [CrossRef]

13. Janssen, A.G.F.; Riedl, T.; Hamwi, S.; Johannes, H.H.; Kowalsky, W. Highly efficient organic tandem solar cells using an improved connecting architecture. *Appl. Phys. Lett.* **2007**, *91*, 073519. [CrossRef]

14. Kim, J.Y.; Lee, K.; Coates, N.E.; Moses, D.; Nguyen, T.Q.; Dante, M.; Heeger, A.J. Efficient tandem polymer solar cells fabricated by all-solution processing. *Science* **2007**, *317*, 222–225. [CrossRef] [PubMed]

15. You, J.; Dou, L.; Yoshimura, K.; Kato, T.; Ohya, K.; Moriarty, T.; Emery, K.; Chen, C.C.; Gao, J.; Li, G.; et al. A polymer tandem solar cell with 10.6% power conversion efficiency. *Nat. Commun.* **2013**, *4*, 1446–1456. [CrossRef] [PubMed]

16. Kim, S.; Kim, C.H.; Lee, S.K.; Jeong, J.H.; Lee, J.; Jin, S.H.; Shin, W.S.; Song, C.E.; Choi, J.H.; Jeong, J.R. Highly efficient uniform ZnO nanostructures for an electron transport layer of inverted organic solar cells. *Chem. Commun.* **2013**, *49*, 6033–6035. [CrossRef] [PubMed]

17. Bouclé, J.; Snaith, H.J.; Greenham, N.C. Simple approach to hybrid polymer/porous metal oxide solar cells from solution-processed ZnO nanocrystals. *J. Phys. Chem. C* **2010**, *114*, 3664–3674. [CrossRef]

18. Zhao, D.; Peng, T.; Lu, L.; Cai, P.; Jiang, P.; Bian, Z. Effect of Annealing temperature on the photoelectrochemical properties of dye-sensitized solar cells made with mesoporous TiO_2 nanoparticles. *J. Phys. Chem. C* **2008**, *112*, 8486–8494. [CrossRef]

19. Takanezawa, K.; Tajima, K.; Hashimoto, K. Efficiency enhancement of polymer photovoltaic devices hybridized with ZnO nanorod arrays by the introduction of a vanadium oxide buffer layer. *Appl. Phys. Lett.* **2008**, *93*, 063308. [CrossRef]

20. Yang, T.; Sun, K.; Liu, X.; Wei, W.; Yu, T.; Gong, X.; Wang, D.; Cao, Y. Zinc oxide nanowire as an electron-extraction layer for broadband polymer photodetectors with an inverted device structure. *J. Phys. Chem. C* **2012**, *116*, 13650–13653. [CrossRef]

21. Baek, W.H.; Seo, I.L.; Yoon, T.S.; Lee, H.H.; Yun, C.M.; Kim, Y.S. Hybrid inverted bulk heterojunction solar cells with nanoimprinted TiO_2 nanopores. *Sol. Energy Mater. Sol. Cells* **2009**, *93*, 1587–1591. [CrossRef]

22. Sekine, N.; Chou, C.H.; Kwan, W.L.; Yang, Y. ZnO nano-ridge structure and its application in inverted polymer solar cell. *Org. Electron.* **2009**, *10*, 1473–1477. [CrossRef]

23. Kim, K.D.; Lim, D.C.; Hu, J.; Kwon, J.D.; Jeong, M.G.; Seo, H.O.; Lee, J.Y.; Lim, K.H.; Lee, K.H.; Jeong, Y.; et al. Surface modification of a ZnO electron-collecting layer using atomic layer deposition to fabricate high-performing inverted organic photovoltaics. *ACS Appl. Mater. Interfaces* **2013**, *5*, 8718–8723. [CrossRef] [PubMed]

24. Lim, D.C.; Shim, H.H.; Kim, K.D.; Seo, H.O.; Lim, J.H.; Jeong, Y.; Kim, Y.D.; Lee, K.H. Spontaneous formation of nanoripples on the surface of ZnO thin films as hole-blocking layer of inverted organic solar cells. *Sol. Energy Mater. Sol. Cells* **2011**, *95*, 3036–3040. [CrossRef]

25. Liao, S.H.; Jhuo, H.J.; Cheng, Y.S.; Chen, S.A. Fullerene derivative-doped zinc oxide nanofilm as the cathode of inverted polymer solar cells with low-bandgap polymer (PTB7-Th) for high performance. *Adv. Mater.* **2013**, *25*, 4766–4771. [CrossRef] [PubMed]

26. Kim, Y.S.; Tai, W.P. Electrical and optical properties of Al-doped ZnO thin films by sol-gel process. *Appl. Surf. Sci.* **2007**, *253*, 4911–4916. [CrossRef]

27. Pita, K.; Baudin, P.; Vu, Q.V.; Aad, R.; Couteau, C.; Lerondel, G. Annealing temperature and environment effects on ZnO nanocrystals embedded in SiO. *Nanoscale Res. Lett.* **2013**, *8*, 517. [CrossRef] [PubMed]

28. Chu, M.C.; You, H.C.; Meena, J.S.; Shieh, S.H.; Shao, C.Y.; Chang, F.C.; Ko, F.H. Facile electroless deposition of zinc oxide ultrathin film for zinc acetate solution-processed transistors. *Int. J. Electrochem. Sci.* **2012**, *7*, 5977–5989.

29. Deegan, R.D.; Bakajin, O.; Dupont, T.F.; Huber, G.; Nagel, S.R.; Witten, T.A. Capillary flow as the cause of ring stains from dried liquid drops. *Nature* **1997**, *389*, 827–829. [CrossRef]

30. Noh, Y.J.; Na, S.L.; Kim, S.S. Inverted polymer solar cells including ZnO electron transport layer fabricated by facile spray pyrolysis. *Sol. Energy Mater. Sol. Cells* **2013**, *117*, 139–144. [CrossRef]

31. Rabani, E.; Reichman, D.R.; Geissler, P.L.; Brus, L.E. Drying-mediated self-assembly of nanoparticles. *Nature* **2003**, *426*, 271–274. [CrossRef]

32. Kim, Y.S.; Tai, W.P.; Shu, S.J. Effect of preheating temperature on structural and optical properties of ZnO thin films by sol-gel process. *Thin Solid Films* **2005**, *491*, 153–160. [CrossRef]

33. Kyaw, A.K.K.; Wang, D.H.; Wynands, D.; Zhang, J.; Nguyen, T.Q.; Bazan, G.C.; Heeger, A.J. Improved light harvesting and improved efficiency by insertion of an optical spacer (ZnO) in solution-processed small-molecule solar cells. *Nano Lett.* **2013**, *13*, 3796–3801. [CrossRef] [PubMed]

34. Lin, S.H.; Lan, S.; Sun, J.Y.; Lin, C.F. Influence of mixed solvent on the morphology of the P3HT:Indene-C60 bisadduct (ICBA) blend film and the performance of inverted polymer solar cells. *Org. Electron.* **2013**, *14*, 26–31. [CrossRef]

35. Cheng, F.; Fang, G.; Fan, X.; Huang, H.; Zheng, Q.; Qin, P.; Lei, H.; Li, Y. Enhancing the performance of P3HTICBA based polymer solar cells using LiF as electron collecting buffer layer and UV-ozone treated MoO_3 as hole collecting buffer layer. *Sol. Energy Mater. Sol. Cells* **2013**, *110*, 63–68. [CrossRef]

36. Yeh, P.N.; Jen, T.H.; Cheng, Y.S.; Chen, S.A. Large active area inverted tandem polymer solar cell with high performance via insertion of subnano-scale silver layer. *Sol. Energy Mater. Sol. Cells* **2014**, *120*, 728–734. [CrossRef]

37. Li, G.; Yao, Y.; Yang, H.; Shrotriya, V.; Yang, G.; Yang, Y. "Solvent Annealing" effect in polymer solar cells based on poly(3-hexylthiophene) and methanofullerenes. *Adv. Funct. Mater.* **2007**, *17*, 1636–1644. [CrossRef]

38. Wang, J.C.; Ren, X.C.; Shi, S.Q.; Leung, C.W.; Chan, P.K.L. Charge accumulation induced S-shape J-V curves in bilayer heterojunction organic solar cells. *Org. Electron.* **2011**, *12*, 880–885. [CrossRef]

39. Liu, F.; Nunzi, J.M. Air stable hybrid inverted tandem solar cell design. *Appl. Phys. Lett.* **2011**, *99*, 063301. [CrossRef]

Publisher's Note: MDPI stays neutral with regard to jurisdictional claims in published maps and institutional affiliations.

Review

Recent Advances of Hierarchical and Sequential Growth of Macromolecular Organic Structures on Surface

Corentin Pigot * and Frédéric Dumur *

Aix Marseille Univ, CNRS, ICR UMR 7273, F-13397 Marseille, France
* Correspondence: pigotcorentin2@gmail.com (C.P.); frederic.dumur@univ-amu.fr (F.D.);
  Tel.: +33-(0)4-91-28-27-48 (F.D.)

Received: 17 January 2019; Accepted: 19 February 2019; Published: 22 February 2019

Abstract: The fabrication of macromolecular organic structures on surfaces is one major concern in materials science. Nanoribbons, linear polymers, and porous nanostructures have gained a lot of interest due to their possible applications ranging from nanotemplates, catalysis, optoelectronics, sensors, or data storage. During decades, supramolecular chemistry has constituted an unavoidable approach for the design of well-organized structures on surfaces displaying a long-range order. Following these initial works, an important milestone has been established with the formation of covalent bonds between molecules. Resulting from this unprecedented approach, various nanostructures of improved thermal and chemical stability compared to those obtained by supramolecular chemistry and displaying unique and unprecedented properties have been developed. However, a major challenge exists: the growth control is very delicate and a thorough understanding of the complex mechanisms governing the on-surface chemistry is still needed. Recently, a new approach consisting in elaborating macromolecular structures by combining consecutive steps has been identified as a promising strategy to elaborate organic structures on surface. By designing precursors with a preprogrammed sequence of reactivity, a hierarchical or a sequential growth of 1D and 2D structures can be realized. In this review, the different reaction combinations used for the design of 1D and 2D structures are reported. To date, eight different sequences of reactions have been examined since 2008, evidencing the intense research activity existing in this field.

Keywords: on-surface reaction; stepwise growth; sequential growth; hierarchical growth; macromolecular organic structures; surface covalent organic framework; nanoribbons; macrocycles; coordination polymers

---

1. Introduction

The discovery of new classes of materials is fundamental to the development of future applications [1] and the possibility to elaborate well-organized structures on surface paves the way towards the design of miniaturized devices and the exploration of new concepts in nanoelectronics [2]. With regards to their exquisite properties, organic compounds are prime candidates because their functionalities are widely variable. During decades, supramolecular chemistry has proved to be a powerful tool to elaborate well-organized and extended structures on surfaces [3–6] (see Figure 1). By applying the fundaments of supramolecular chemistry, the monomers which constitute the elemental building-blocks for the construction of two-dimensional structures could be arranged precisely, with a long-range order by mean of secondary interactions (van der Waals interactions, hydrogen bonds, etc.).

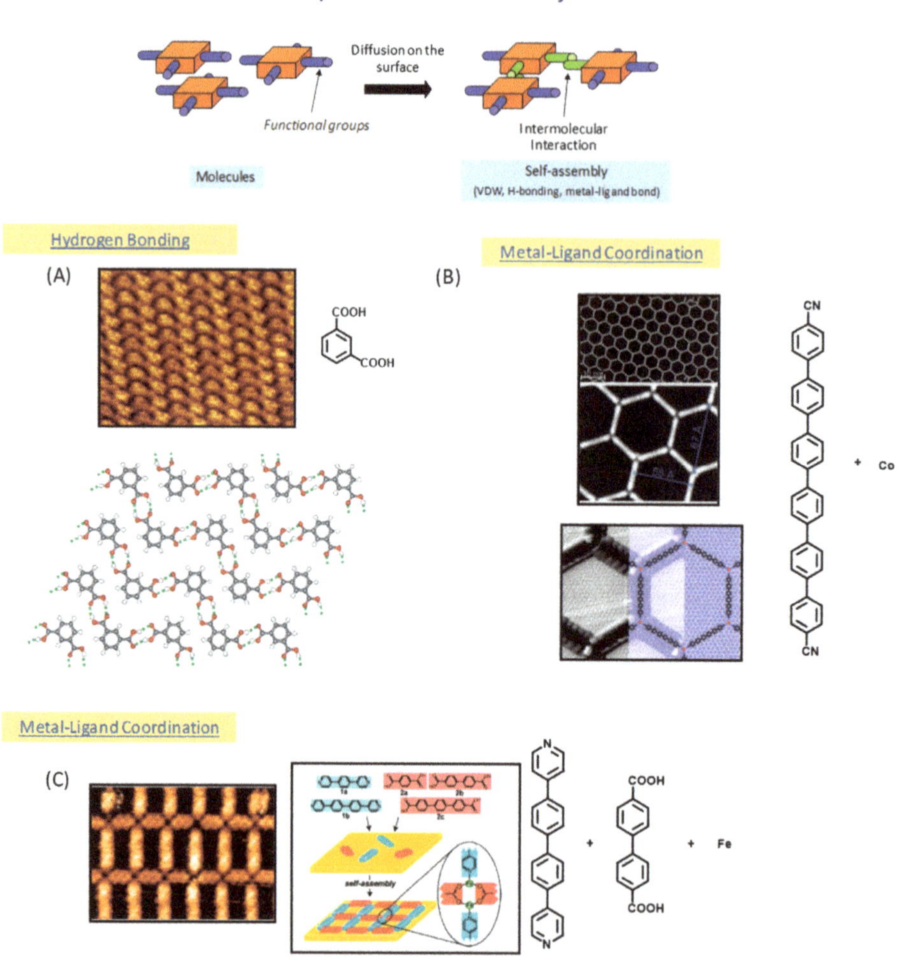

Figure 1. Principle of the supramolecular self-assembly. (**A**) Scanning tunneling microscope (STM) image of isophthalic acid on highly ordered pyrolytic graphite (HOPG) and picture of the molecular mechanics simulation. Adapted with permission from Lackinger et al. [3]. Copyright 2004 American Chemical Society. (**B**) STM images of the metal–organic supramolecular structure formed with cobalt on Ag(111) and superimposition of a ball and stick model onto the STM image. Reprinted with permission from Kühne et al. [5]. Copyright 2009 American Chemical Society. (**C**) STM image of the iron coordination network and schematic illustration of the complementary ligands coordinated to iron, Langner et al. [6]. Copyright 2007 National Academy of Sciences.

Parallel to this, due to the noncovalent character of the bonds created between tectons [7–10], defects formed during the initial growth of the supramolecular phase can be partially or totally repaired, self-healing consisting in a thermal post-treatment of the surface. Resulting from the possibility to suppress defects, networks perfectly ordered and extending over tens of square nanometers could be prepared (see Figure 1). By chemical engineering of the molecular tectons, the design of complex structures with precise shapes, compositions and functionalities is thus rendered possible. The large number of published studies and review articles dealing with supramolecular self-assemblies at surfaces demonstrates it as a mature field. However, the low thermal stability and the lack of electronic conjugation often evidenced in these assemblies drastically limit the applicability of these systems [11–13]. To overcome these drawbacks, an efficient strategy consists in generating covalent bonds between tectons, but it requires the monomers to be conveniently functionalized (see Figure 2) [14–20].

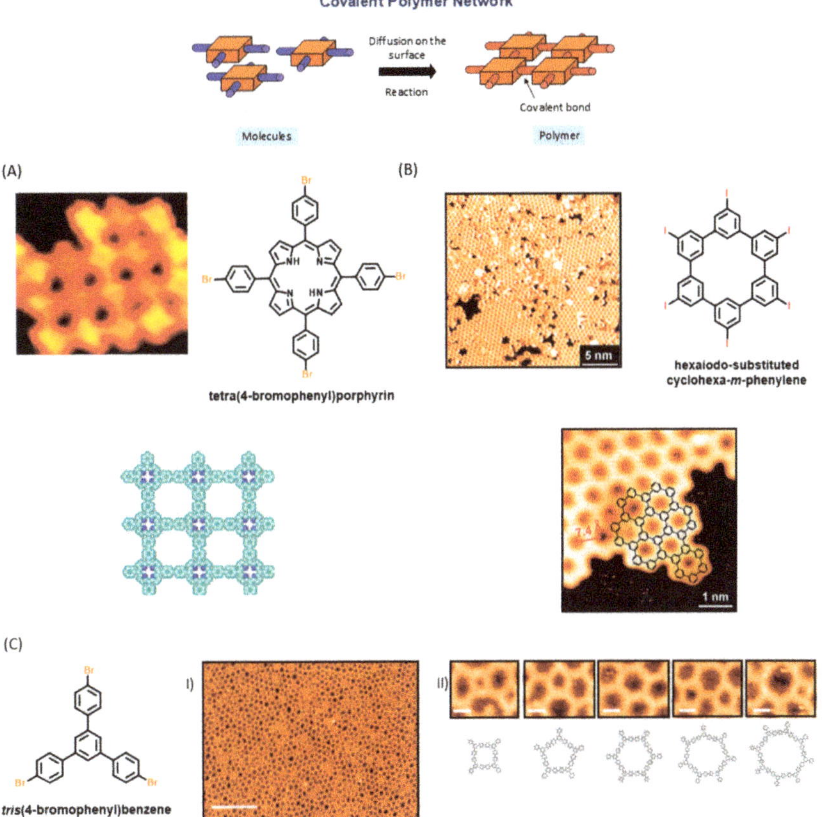

Figure 2. Principles of the formation of covalent polymer network on surface. (**A**) STM image of the polymerization of tetra(4-bromophenyl)porphyrin on Au(111). Reprinted by permission from Springer Customer Service Center GmbH: Springer Nature, Nature Nanotechnology, Grill et al. [18], 2007. (**B**) STM image of the polymer formed with hexaiodo-substituted cyclohexa-*m*-phenylene on Ag(111). Reproduced from [19] with permission from The Royal Society of Chemistry. (**3**) STM image of the polymer formed with *tris*(4-bromophenyl)benzene on Au(111). Scale bars: (**I**) 190 Å, (**II**) 14 Å. Adapted from [20] with permission from The Royal Society of Chemistry

In this field, formation of covalent bonds between tectons is not limited to two-dimensional (2D) structures and one-dimensional (1D) structures have also been the purpose of numerous studies. Due to the remarkable robustness resulting from the formation of covalent bonds between molecular subunits, the excellent thermal, mechanical and chemical stability, and the possibility to create fully conjugated structures with the desired functionalities, a great impact of these covalent structures in nanoelectronics can already be anticipated, irrespective of their shapes, sizes, and lengths [21]. As far as 2D polymers are concerned, the most basic organic material with a strictly 2D character is graphene, which is composed solely of carbon atoms, and which is attracting considerable interest nowadays as illustrated, in particular, by year 2010's Nobel Prize award in Physics. Under intense investigations by various groups in the world, the properties of graphene appear simply exceptional in terms of basic science as well as for practical applications in electronic devices [22]. However, applications such as solar cells [23] or field-effect transistors [24] require a high level of control on its doping level to pertain its properties in devices, for example by introducing chemical changes [25]. Indeed, graphene's properties are very sensitive to chemical modifications [26], which can in principle be extended infinitely to form a well-ordered sheet composed of various heteroatoms. The development of 2D polymers is thus a meeting point where the extraordinary properties of a 2D material such as graphene can be combined with the powerful knowledge of supramolecular and synthesis chemistry to create a new class of materials of quasi-infinite structural and functional diversity [27–29]. However, growth control of these structures remains very delicate and there is still a lot of work to be done to improve to optimize the on-surface synthesis.

At present, possibilities offered by organic chemistry to form covalent bonds between molecular tectons have only been scarcely explored by physicists with regards to the number of existing coupling modes [30,31] and the reactions of cyclodehydrogenation [32,33], dehydration of boronic acids [34–39], esterification of boronic acids [40], Bergman cyclization [41], Glaser coupling [42,43], Wurtz reaction [44], dehydrogenative coupling of terminal alkenes [45], dehydrogenative coupling of porphyrines [46], aryl–aryl coupling via a C–H activation [47], formation of triple bonds by coupling of trichloromethyl groups [48], cyclotrimerization of acetyls [49], formation of imines [50–58], esters [59], imides [60,61], amides [62,63], or the Ullmann coupling between halogenated aromatic rings [18,64] can be cited as examples of reactions already examined on surface (see Figure 3). Parallel to known reactions, unprecedented coupling modes were also discovered in the context of the surface-mediated reactions, as exemplified with the oxidative coupling of activated methylene groups [65–67], the polymerization of alkanes [68], or the aryl–aryl coupling of none halogenated polyaromatics [47] or porphyrins [69], reactions that have never been reported in solution phase chemistry [70].

In the search for nanostructures of high thermal, mechanical, and chemical stability, a major breakthrough has thus been achieved with the demonstration that the formation of covalent bonds between tectons was possible [18–20]. However, rapidly, pioneering works on surface covalent organic frameworks (sCOFs) have evidenced the difficulties of the polymer growth. Notably, due to the formation of covalent bonds and the low-dimensional environment of the on-surface syntheses, possibilities of defect self-healing remain limited, and can even be considered as being precluded for certain reactions such as Ullmann coupling [64], Wurtz coupling [45], or Bergman cyclization [41]. In these conditions, the resulting macromolecular organic structures formed with these reactions are either poorly ordered and limited in terms of 1D or 2D growth. Conversely, all reactions based on a dehydration process are reversible and the possibility of self-healing remains possible, as exemplified with imines [55,57] or boronic esters [40]. To increase the size and the regularity of the nanodomains made of 1D or 2D structures, novel and more efficient chemistries or synthetic approaches offering a better control of the growth conditions are highly desired and will guide the future development of this emerging technology. Concerning growth control, the best candidates to examine this point are undoubtedly 2D structures requiring the covalent linkage to be carried out in the two dimensions and on the basis of the different works reported in the literature, some conclusions can be deduded. Notably, the simultaneous presence of numerous identical and reactive functional groups onto molecular tectons was identified as favoring the formation of numerous defects, the probability for the monomer to react in an inappropriate orientation being drastically increased [71]. This drawback was notably evidenced with diboronic acids during the formation of 2D polymers. Due to an insufficient space to promote the diffusion and the rearrangement of 1,4-benzenediboronic acid (BDBA) onto the surface, various pore shapes could be found on the surface while using this molecular building block (see Figure 4) [29].

Another relevant example of a drawback induced by the presence of numerous reactive functional groups onto the same monomer is provided with hexaiodo-substituted cyclohexa-*m*-phenylene [19]. In this last case, during the thermal deposition of the monomer onto a Ag(111) surface, an undesired dehalogenation reaction occurred, promoting the formation of numerous defects on the surface (see Figure 4).

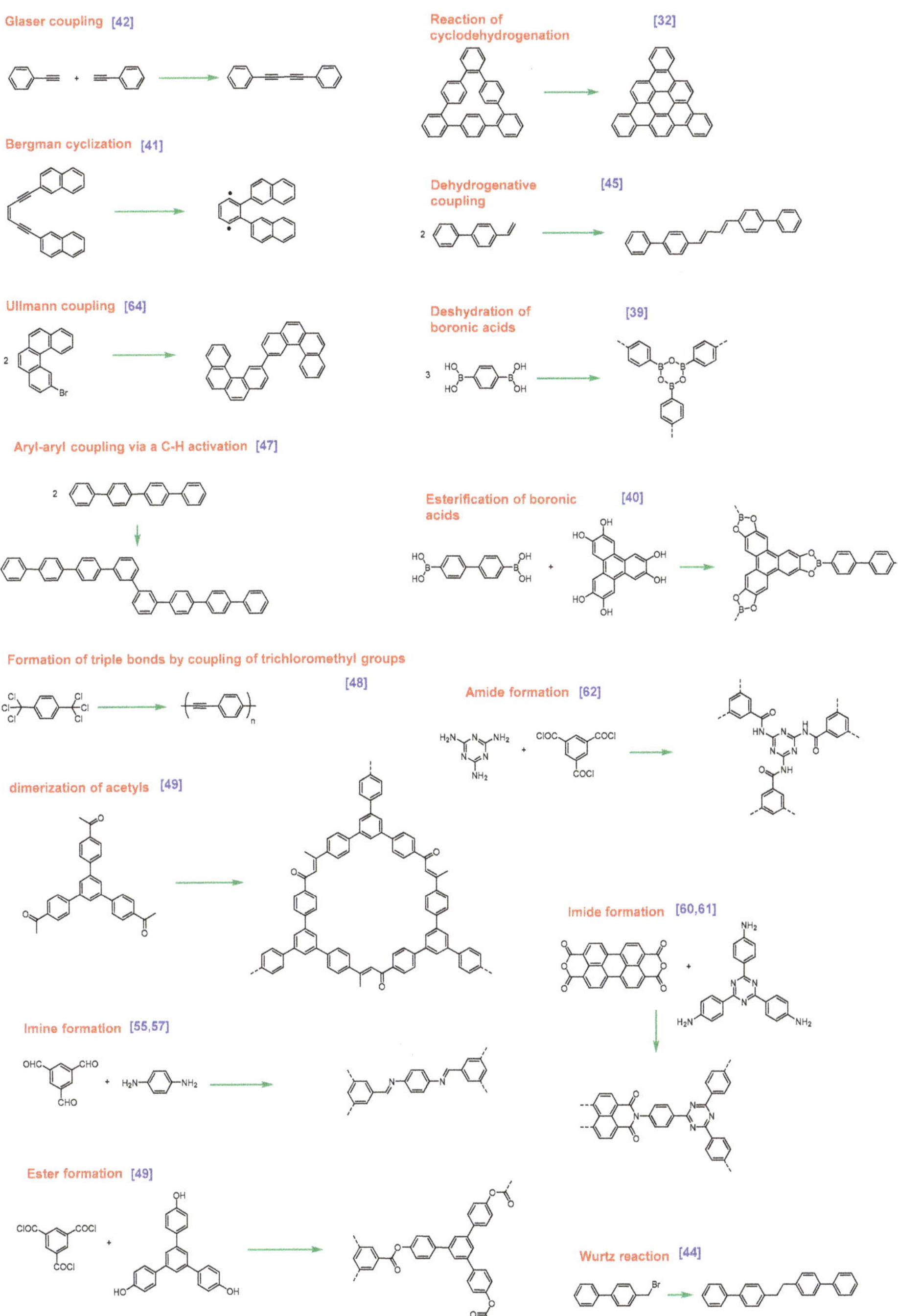

Figure 3. Examples of common on-surface reactions reported in the literature.

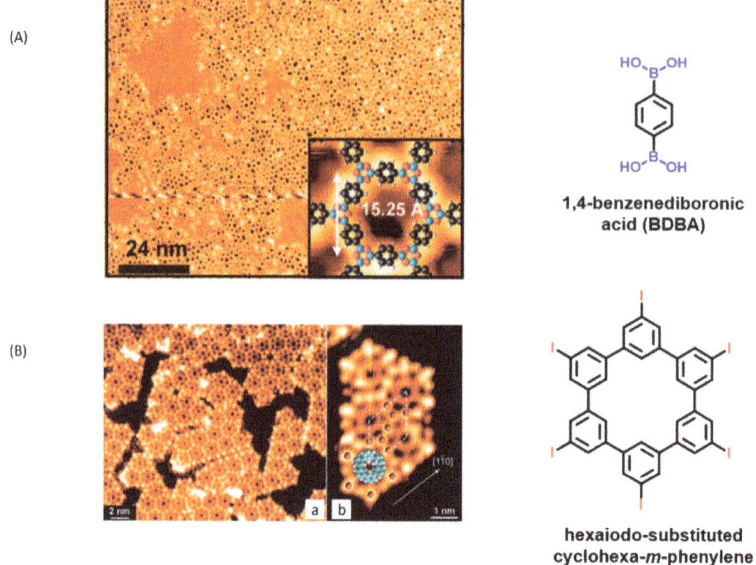

Figure 4. (**A**) Disorder exists during the polymer growth of 1,4-benzenediboronic acid (BDBA) on Au(111). Reprinted with permission from Zwaneveld et al. [39]. Copyright 2008 American Chemical Society. (**B**) Loss of iodine atoms during the thermal deposition of hexaiodo-substituted cyclohexa-*m*-phenylene on a Ag(111) surface kept at room temperature (red spheres correspond to iodine atoms). (**Ba**) STM image of the macromolecular structure on Ag(111); (**Bb**) STM image, magnification of a nanodomain. Adapted from [19] with permission from The Royal Society of Chemistry

To circumvent these different problems, alternative procedures to construct nanoporous structures have been actively researched. In most of the works reported in the literature, molecular tectons only possess a single type of functional groups, inducing a disorder during polymerization as several functional groups of a same monomer are simultaneously involved in reactions. Loss of functional groups is also occasionally observed.

Recently, the development of monomers bearing different types of functional groups activable in different reaction conditions has emerged as an effective way to address the defect issue. Typically, monomers developed for this purpose possess two (or more) functional groups that can be activated separately. More precisely, the dormant sites can be activated in a predetermined order, giving rise to a sequential growth of the final network. The reactive sites are also selected to be compatible so that no reaction should occur between the different functional groups. Such reactions that do not interfere each other are typically named orthogonal reactions [72–76]. Parallel to the sequential growth of polymers, the hierarchical growth of sCOF can also be employed but requires the monomers to be substituted adequately. As the main difference between the sequential and the hierarchical growths of macromolecular architectures, two different types of functional groups are involved contrarily to the hierarchical growth for which the same types of functional groups can be used, but can be activated in different reaction conditions. In this field, the most representative example concerns aromatic precursors bearing two different types of halogens, namely bromines and iodines. Even if an Ullmann coupling can occur with the two types of halogens, selectivity will arise from the difference of reactivity between iodines and bromines. By perfectly controlling the reaction temperature, Ullmann coupling will occur for the iodinated sites prior to the brominated ones. Using this stepwise strategy (hierarchical or sequential), nanoporous architectures extending over tens of square nanometers can be obtained, resulting from an advanced polymerization process at each step. During the last decade, major breakthroughs have been achieved in the hierarchical and sequential growth of 2D polymers. However, this synthetic approach is not limited to 2D structures and 1D structures such as ribbons, macrocycles, and linear polymers have also been prepared using this innovative strategy. Indeed, the contribution of 1D and 2D structures in the comprehension of the underlying mechanisms of polymerization are comparable, especially due to the fact the reactions involved in the sequential or hierarchical growth of 1D and 2D organic structures are different at only one exception. Indeed, a recent work succeeded to combine the Ullmann/aromatization sequential growth strategy typically used for the elaboration of 1D structures to the deshydrogenative cross-coupling of aromatics, enabling the bottom-up synthesis of two dimensional graphene-like structures [77]. To image these structures and the different growth steps of the macromolecular organic structures, scanning tunneling microscopy (STM) is a remarkable technique combining the atomic resolution with the ability to probe the local electronic structure [78]. To get a deeper insight into the chemical structures of the macromolecular organic structures, STM can be combined with other surface analysis methods such as Raman spectroscopy, photoemission, high resolution electron energy loss spectroscopy (HREELS), or synchrotron radiation analyses [79,80]. The combination of these different techniques is of crucial importance, especially to characterize the multistep growth of macromolecular organic structures. Before continuing, a distinction should be made between "self-assembly", which is commonly used to evoke the formation of both supramolecular and covalent phases on-surface, and "chemically released diffusion". Indeed, numerous published or being published articles comprise the terminology "self-assembly" whereas the right term is "chemically released diffusion". Indeed, "self-assembly" refers to the self-organization of molecular tectons on surface without necessarily involving a chemical modification of the molecular tectons and this terminology is well-adapted for supramolecular phases for which the cohesion is ensured by weak intermolecular interactions between elemental building blocks. Contrastingly, chemically released diffusion refers to mass transfer occurring in polycondensates due to the permanent exchange of fragments existing between molecular segments under growth (i.e., via transreactions). This concept developed in the early 1980s by Prof. Stoyko Fakirov is more adapted to describe the formation of covalent phases on surface, the chemical composition of the polymer under growth continuously evolving by exchange reactions [81].

In this review, an overview of the recent advances concerning the sequential or hierarchical growth of macromolecular organic structures on surface in ultrahigh vacuum (UHV) environment is presented. As compared to their one-step counterparts, multistep formation of organic structures on surface is less documented in the literature. However, elaboration of covalent bonds in multistep procedures offers a unique opportunity to guide the growth of the macromolecular architecture and to control the kinetic of reaction at each individual step. This hierarchical and sequential approach is of prime importance for the regularity of 2D polymers on surface, but also for the preparation of 1D structures as the efficacy of the covalent couplings will determine the length of the final structures. At present, most of the reaction combinations used for the sequential growth of 1D and 2D structures are different. To date, eight different combinations of reactions have been examined in the literature and these strategies are presented in this review.

2. Coupling Modes Used for the Design of 2D Covalent Networks

2.1. Hierarchical Growth of Macromolecular Architectures Based on Successive Ullmann Couplings

On-surface chemistry is an emerging field of research and optimization of the reaction conditions to elaborate well-defined nanostructures constitutes the major focus of researchers. Since the first reports mentioning the elaboration of sCOFs, Ullmann coupling of halogenated aromatic rings [31,82] have undoubtedly been one of the most commonly used coupling mode. As mentioned previously, the sequential growth of covalent networks is mainly based on the attachment of two different types of functional groups onto the same monomer. Deviating from this situation, the reactivity of halogens is well known to differ when two different types of halogens are attached to the same molecules. Numerous reactions in solution phase chemistry are based on these differences of reactivity between halogens and this specificity was transposed to on-surface synthesis. In this case, Ullmann coupling can be selectively activated by a careful control of the reaction temperature and fully conjugated 2D polymers can be prepared. This situation typically constitutes a hierarchical growth of 2D structures. In this field, the first report mentioning a hierarchical approach was published in 2012 by Grill et al. [83]. To achieve a complete control during the hierarchical growth, iodine and bromine atoms were introduced onto tetraphenyl porphyrin to give 5,15-*bis*(4′-bromophenyl)-10,20-*bis*(4′-iodophenyl)porphyrin (*trans*-Br$_2$I$_2$TPP) (see Figure 5A).

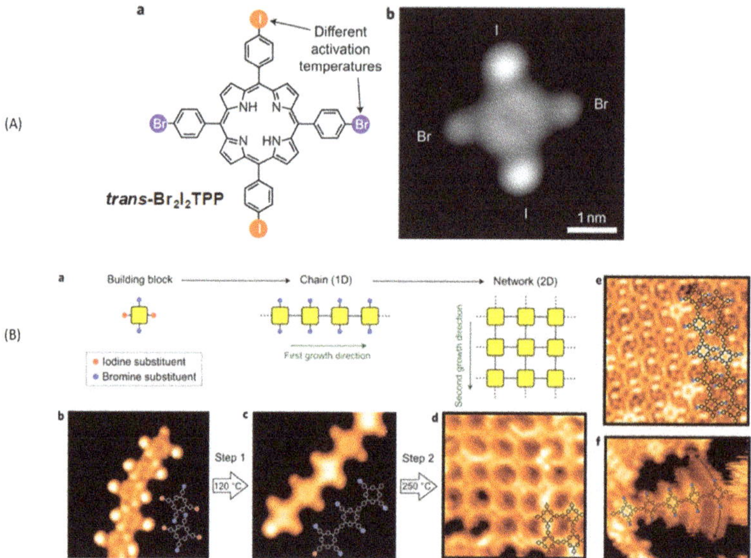

Figure 5. Sequential growth of a 2D architecture with 5,15-*bis*(4′-bromophenyl)-10,20-*bis*(4′-iodophenyl) porphyrin (*trans*-Br$_2$I$_2$TPP). (**A**) Chemical structure of the precursor. (**B**) Two-step polymer growth. Reprinted by permission from Macmillan Publishers Ltd: Nature Chemistry from [83], copyright 2012.

The bond dissociation energy (BDE) of these carbon-halogen bonds are different and determined in the gas phase as being 336 and 272 kJ/mol for the C–Br and C–I bonds, respectively. To anticipate the reactivity of monomers on surface, determination of BDEs in the gas phase are not sufficient. Indeed, the role of the metal surface is crucial (metal choice, crystallographic plane, etc.) and the possibility for the monomers to interact with the substrates are not taken into account in the theoretical calculations. Based on previous works devoted to the Ullmann coupling of halogenated precursors on surface, a scale of reactivity of the carbon–halogen bond could be established, taking in consideration these different parameters. Thus, the C–I bond is well-known to spontaneously cleave on copper, silver, and gold [28,84]. Conversely, the C–Br bond can fully cleave on copper [85,86], partially on silver [87,88], and remains intact on gold [20] in the absence of thermal activation. In this context, the combination of iodine and bromine halogens on *trans*-Br$_2$I$_2$TPP was appropriate for works carried out on gold. Precisely, the deposition of *trans*-Br$_2$I$_2$TPP was examined on Au(111) substrates. After deposition of *trans*-Br$_2$I$_2$TPP at −93 °C, which let the monomer intact, the sequential growth could be activated by heating first the surface at 120 °C and then 250 °C. Here again, the well-separated temperature windows enabled to perfectly control the two-step process and to avoid the possibility of competing Ullmann couplings. As shown in the Figure 5B, the first step produced linear chains of porphyrins by activation of the iodine sites, without forming lateral connection between the different chains.

By activating the second growth direction at a higher temperature, a 2D architecture was obtained, resulting from the crosslinking of the different chains. Regular structures extending on surfaces of area bigger than 10×10 nm^2 could be imaged. Catalytic activity of the gold surface in this second step was clearly evidenced as the C–Br bond dissociation could be obtained at 200 °C on the gold surface, below the temperature required to get a bromine dissociation in the evaporator (300 °C) [83].

The possibility to design more complex structures was examined by copolymerizing *trans*-Br$_2$I$_2$TPP with dibromoterfluorene (DBTF) on Au(111) surfaces (see Figure 6A). Thanks to the sequential procedure, even if both types of molecules are deposited onto the metal surface, heating of the surface at 250 °C first induced the formation of porphyrin chains by activation of the iodine sites (only 2% of undesired reaction of DBTF with *trans*-Br$_2$I$_2$TPP was detected) followed in second step by the carbon–bromine dissociation resulting in the covalent coupling of the porphyrin chains with DBTF. Chains with length ranging between 15 and 20 nm could be imaged. Even if the lateral connection of adjacent porphyrins was detected on the STM images (see Figure 6B,C), 70% of the carbon–carbon bonds resulted from the reaction between the bromine sites of porphyrin and DBTF, opening the way towards the design of complex architectures. Following this work, the design of 2D polymers with 1,3-*bis*(*p*-bromophenyl)-5-(*p*-iodophenyl)benzene (BIB) was examined (see Figure 7A) [89]. Here again, the two steps are characterized by a significant difference of their activation temperatures since the first step (activation of the iodine sites of BIB) could be realized at room temperature whereas an annealing at 185 °C was required to initiate the second step (activation of the bromine sites of BIB). As anticipated, deposition of BIB on Au(111) surfaces at room temperature resulted in the dimerization of BIB and the formation of 3,3′′′,5,5′-tetra(*p*-bromophenyl)-1,1′′,1′′:4′′,1′′′-quaterphenyl (TBQ). STM images of the supramolecular phase revealed the TBQ units to be linearly aligned in parallel rows, with iodine atoms standing between the TBQ motifs.

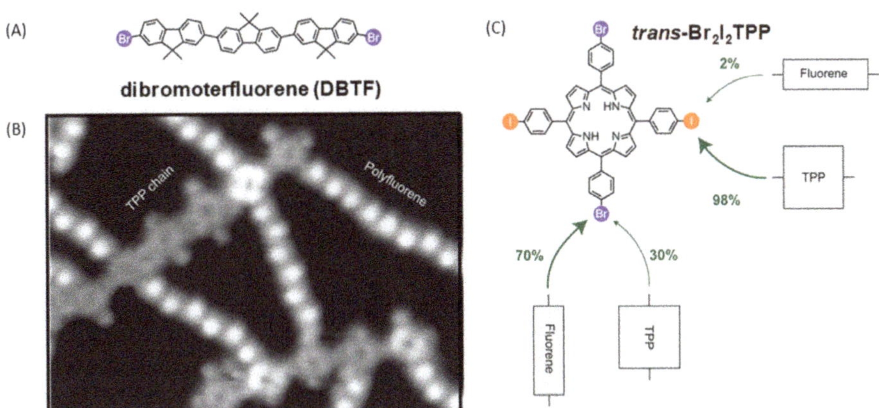

Figure 6. (**A**) Chemical structure of dibromoterfluorene (DBTF). (**B**) STM topograph of the copolymerization of 5,15-*bis*(4′-bromophenyl)-10,20-*bis*(4′-iodophenyl)porphyrin (*trans*-Br$_2$I$_2$TPP) with DBTF (13×18 nm^2). (**C**) Statistical analyses of the porphyrin/TBTF connection. Reprinted by permission from Macmillan Publishers Ltd: Nature Chemistry from [83], copyright 2012.

The influence of the reaction temperatures (185, 250 and 375 °C) and the heating rates (between 6.3 and 8.9 °C/min) on the second step were examined. Irrespective of these two parameters, a similar disorder accompanied with an incomplete coverage of the surface and the formation of pores could be found on all STM images, of scanning size 60×60 nm^2 for the bigger STM images (see Figure 7B). Complex networks based on highly branched structures could be imaged. As interesting findings, hexagons and opened pores were mainly observed upon annealing the surface at 185 °C whereas squares and octagons were more frequently observed at 250 and 375 °C (see Figure 7C,D).

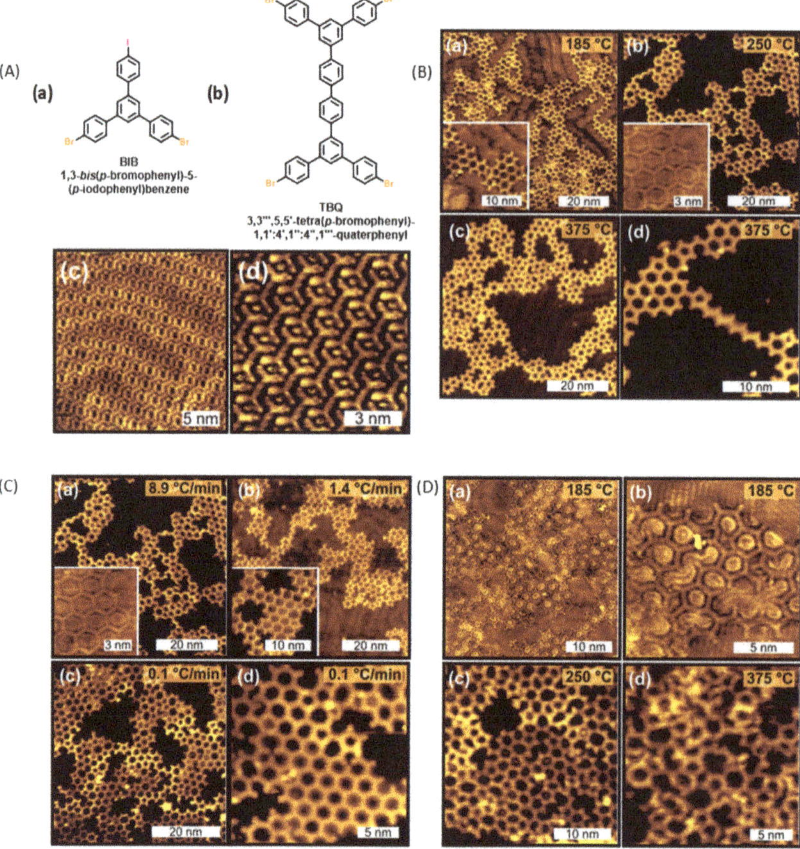

Figure 7. (**Aa,b**) Chemical structures of 1,3-*bis*(*p*-bromophenyl)-5-(*p*-iodophenyl)benzene (BIB) and TBQ. (**Ac,d**) STM images of supramolecular self-assembly of 3,3‴,5,5′-tetra(*p*-bromophenyl)-1,1′:4′,1″:4″,1‴-quaterphenyl (TBQ) on Au(111) surface. (**B**) STM images of the 2D polymers obtained by annealing the surface at 185, 250, and 375 °C. (**C**) STM images of the 2D polymers obtained by annealing the surface at 250 °C at different heating rates. (**D**) STM images of the 2D polymers obtained by direct polymerization at different temperatures. Reprinted with permission from [89]. Copyright 2014 American Chemical Society.

Parallel to the hierarchical polymerization, the possibility of creating regular 2D polymers in one step, by direct polymerization, was examined. In this second approach, temperature of the substrates is determinant since 2D polymers of low quality were obtained for surfaces held at 185 and 375 °C whereas regular hexagonal pores were obtained at 250 °C (see Figure 7D). Almost defect-free polymers extending over surfaces of area bigger than 30×30 nm^2 were detected. At such high temperatures, the deposition rate is another parameter affecting the network quality and reduction of the rate resulted in a broader distribution of polygons going from squares to octagons. From these different experiments, it could be concluded that at low temperature, the mobility of molecules was insufficient and the cleavage of C–Br bonds incomplete so that it could account for the large amount of opened pores. Conversely, at high temperature, the probability to generate pores deviating from the ideal hexagons was greatly enhanced. Finally, comparison of the direct and the hierarchical polymerization revealed the first approach to produce a denser polymeric network even if the density of defects was similar in the two cases. Two years later, other authors examined the covalent coupling of 4-bromo-4″-chloro-5′-(4-chlorophenyl)-1,1′:3′,1′-terphenyl (BCCTP) which differs from BIB by the choice of the halogens (see Figure 8A) [90].

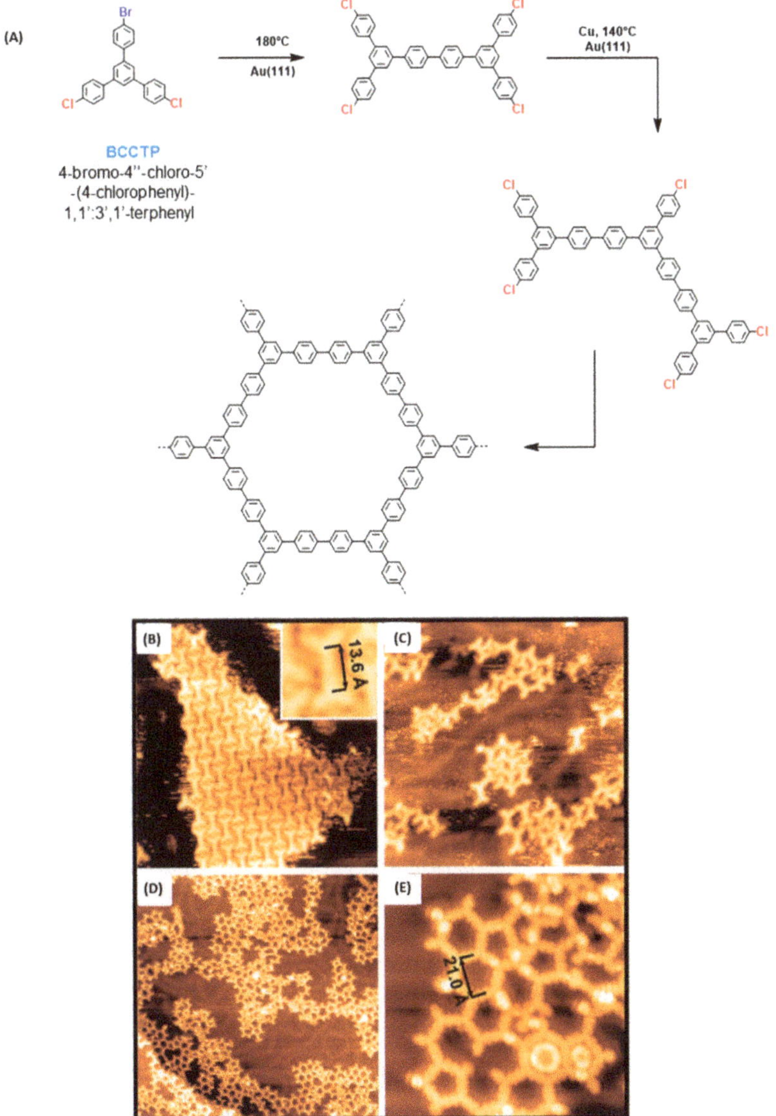

Figure 8. (**A**) Reaction pathway involved in the formation of hexagonal pores with BCCTP. (**B**) Evaporation of BCCTP on Au(111) substrates (26 nm × 26 nm with inset 3 nm × 3 nm). (**C**) Evaporation of the copper catalyst on a Au surface heated at 140 °C (26 nm × 26 nm). (**D,E**) STM images of the Au(111) substrate after annealing the surface at 280 °C at different magnifications (65 nm × 65 nm for (**D**) and 14 nm × 14 nm for (**E**)). Adapted from [90] with permission from The Royal Society of Chemistry.

In this last case and due to the low reactivity of the chlorine atoms, presence of Cu catalyst was required to induce a surface-assisted Ullmann coupling. This is among the first example of Ullmann coupling where a catalyst was introduced to enforce the reaction to occur. As anticipated, and due to the inexistent reactivity of the chlorine atoms, the first step furnished on Au(111) substrates a supramolecular arrangement comparable to that obtained with BIB, with the formation of BCCTP dimers linearly aligned in parallel rows, with bromine atom standing between dimers (see Figure 8B). By evaporating the Cu catalyst on the Au(111) surface kept at 140 °C, the second coupling mode could be activated, enabling the formation of oligomers (see Figure 8C). However, if the polymerization of BCCTP could starts at 140 °C, this latter could only be ended by annealing the surface at 280 °C for 30 minutes (see Figure 8D,E). Once again, a large diversity of polygons was detected on the surface, going from pentagons to heptagons, demonstrating the difficulty to elaborate regular pores. For comparison, a worse result was obtained while depositing BCCTP on Cu(111). Indeed, no control of the polymerization process was possible on this metal surface. Considering that the sequential polymerization of BIB and BCCTP should produce exactly the same final polymers, that the crystallographic plane and the metal chosen for the two studies are the same, and with regards to the size of the STM images (35×35 nm^2 for the polymerization of BIB (see Figure 7(Dc)), 65×35 nm^2 for the polymerization of BCCTP (see Figure 8D), the most regular structures are clearly obtained with BIB which is also the monomer exhibiting the best design to get an advanced polymerization process. In 2017, an interesting study demonstrated the possibility to elaborate a hierarchical growth of 2D polymers with a precursor only bearing one type of halogens [91]. This strategy is quite unexpected considering that the six bromine atoms of 1,3,5-*tris*(3,5-dibromophenyl)benzene (TDBB) could theoretically cleave simultaneously upon thermal activation.

In fact, authors demonstrated this precursor to possibly give rise to two different types of dimers, depending of the coupling modes (type-A and type-B, see Figure 9A–C).

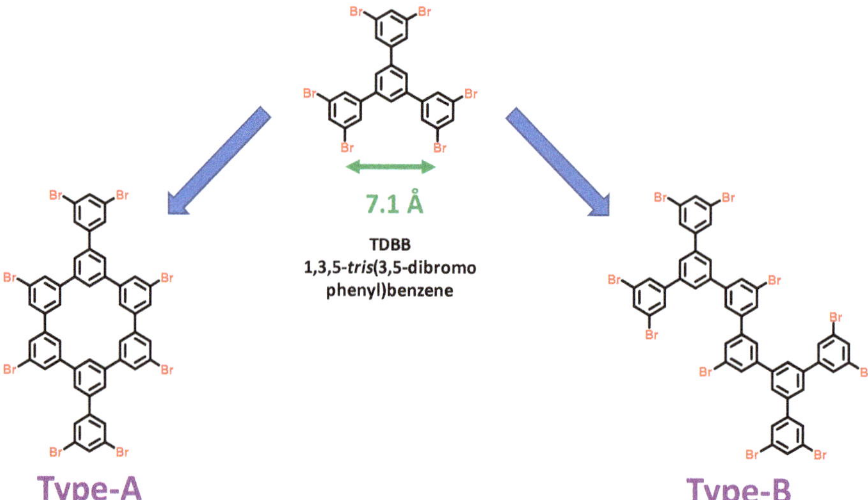

Figure 9. Scheme of the 1,3,5-*tris*(3,5-dibromophenyl)benzene (TDBB) dimer building block with the two-covalent-bond dimer (type-A) and the one-covalent-bond dimer (type-B).

In this work, we succeeded in developing temperature-dependent engineering of 2D polymers. Notably, by annealing at 145 °C the supramolecular phase obtained after evaporation of TDBB on Au(111) surfaces—only type-B dimers could be found on the surface—resulting from the formation of a unique type of C–C bond between TDBB (see Figure 10A). Interestingly, rows of type-B dimers were found to be separated from each other by nanodomains only composed of unreacted TDBB. On the opposite, by annealing the supramolecular phase at 170 °C, oligomers of TDBB formed, resulting from multiple type-B couplings (see Figure 10B).

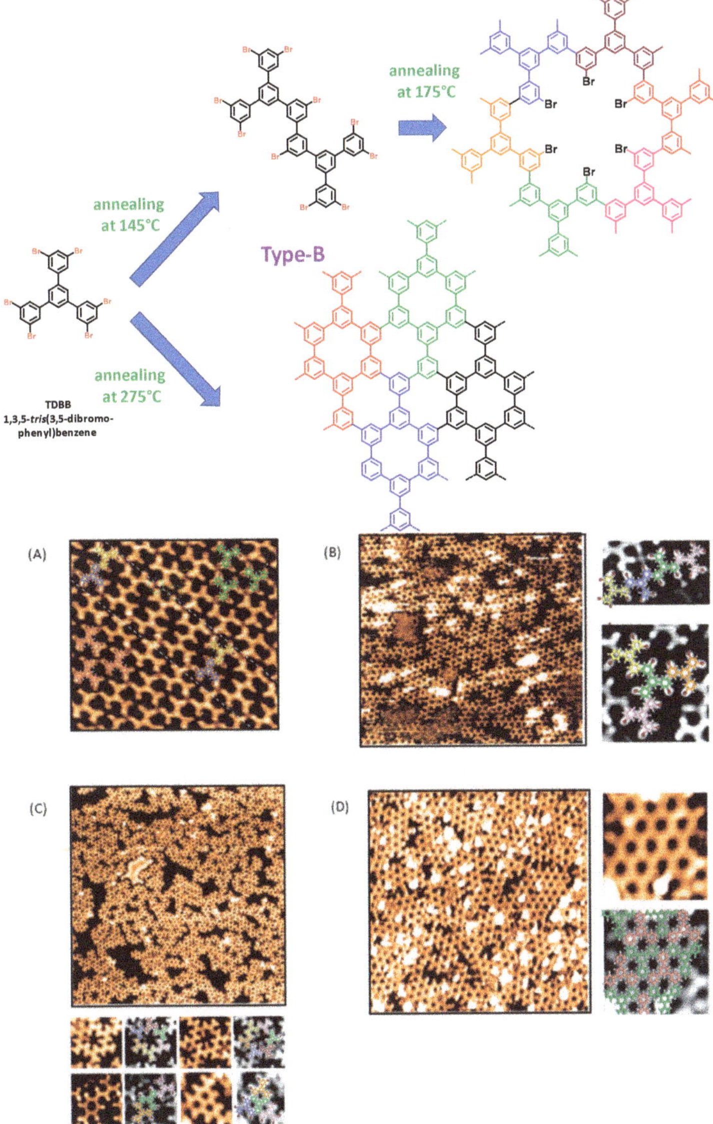

Figure 10. Sequential growth of 2D polymers based on 1,3,5-*tris*(3,5-dibromophenyl)benzene (TDBB). (**A**) STM images of the Au(111) surface after annealing at 145 °C (9 nm × 9 nm). (**B**) STM images of the Au(111) surface after annealing at 170 °C (25 nm × 25 nm). (**C**) STM images of the Au(111) surface after annealing at 175 °C (36 nm × 36 nm and 5 nm × 5 nm (bottom left), 3 nm × 3 nm (bottom right)). (**D**) STM images of the Au(111) surface after annealing at 275 °C (left: 20 nm × 20 nm, right: 3 nm × 3 nm). Adapted with permission from Peyrot at al. [91]. Copyright 2017 American Chemical Society.

Locally, star-shaped tetramers could even be found, still resulting from type-B couplings. Regarding their specificity, these tetramers are issued from the covalent linkage of a central TDBB unit surrounded by three peripheral TDBB. By annealing the supramolecular phase at 175 °C (i.e., at a temperature only higher of 5 °C compared to the previous experiments), the coexistence of different structures could be found, as exemplified in the Figure 10C. Notably, cyclic structures composed of six molecules connected by type-B linkages could be found all over the surface. More scarcely, structures based on type-A couplings could also be found. However, irrespective of the type of linkage, the formation of a dense polymeric structure with dimensions bigger than 40×40 nm^2 (i.e., the size of the STM images) can be detected on Figure 10D, demonstrating the efficient of the Ullmann coupling on Au(111) surfaces. Finally, annealing of the supramolecular phase at 275 °C enabled to form a densely packed 2D networks with hexagonal pores exclusively based on type-A linkages (see Figure 10D). Regularity of the final polymers could be evidenced with a scan area size larger than 20×20 nm^2.

From these different experiments, the sequential growth of the 2D polymers could be deduced. Thus, annealing of the supramolecular phase resulted in first step in the dimerization of TDBB via a single type-B connection. Starting at 170–175 °C, the covalent linkage of three (or more) TDBB units by mean of two type-B connections occurs, giving rise to the formation of linear chains. By increasing the temperature until 275 °C, connections between chains can be initiated, inducing the formation of type-A linkages between molecular tectons. In turn, due to its stepwise growth, a porous network exhibiting a low defect density can be obtained.

2.2. Sequential Growth Based on the Boroxine Formation/Ullmann Coupling Combination

Since the first reports mentioning the elaboration of sCOFs, condensation of boronic acids to form boroxines [92,93] and the Ullmann coupling of halogenated aromatic rings [31,82] have been widely used as coupling modes. Considering that these two on-surface reactions are now a mature field of research with regards to the number of publications, a combination of these two reactions was logical to create 2D covalent networks. The first report mentioning such a combination was published in 2011 [94]. First of all, orthogonality of the two reactions is obvious and no side-reactions are expected to occur at each step. 3,5-Dibromophenylboronic acid (DBPBA) was selected as the molecular tecton and choice of the boronic functional group for DBPBA was dictated by the fact that the boroxine cycle that results from the trimerization process of boronic acids exhibits the same symmetry and almost the same size than a phenyl ring, while sharing its planarity. Under UHV and upon deposition onto Ag(111) substrates, dehydration of DBPBA could be initiated at low temperature produced the unreactive 1,3,5-*tris*(3′,5′-dibromophenyl)boroxine (TDBPB) (see Figure 11). It has to be noticed that the dehydration reaction of boronic acids possesses its own specificity as this reaction does not require the presence of a catalyst and directly occurs during heating in the evaporator. Indeed, upon heating, condensation of boronic acids is preferred over sublimation. In situ formation of TDBPB was confirmed by Raman spectroscopy and by NMR analyses. An additional proof of the formation of trimers was obtained by sublimation of DBPBA onto graphite(001) substrates; only the presence of TDBPB was detected on the STM topographs. No trace of DBPBA was observed on the surface, demonstrating that the cyclodehydration of boronic acids was quantitative in the evaporator. Stability of the supramolecular phase is ensured by electrostatic interactions between halogens, assembled into trigonal cyclic arrangements (see Figure 11A). Interestingly, annealing of the graphite(001) substrates at 200 °C only resulted in the desorption of the molecules whereas annealing of the Ag(111) substrates at lower temperature (130 °C) furnished the targeted polymer by a surface-catalyzed reaction. The higher ability of silver to interact with organic molecules than graphite is a general trend reported in numerous works concerning the on-surface synthesis and the better adsorption of TDBPB on Ag(111) substrates over graphite substrates confirms this trend. Due to the specific substitution of DBPBA (and therefore of TDBPB), the coexistence of two hexagonal arrangements (A-type and B-type) could be potentially observed on the surface (see Figure 11A,B). However, the periodicity of the final network, the distance between cycles, and the symmetry of the overall system revealed the exclusive formation of A-type arrangements. As a drawback, the final polymer lacks long-range order and only small nanodomains were detected on the surface (see Figure 11C). Indeed, nanodomains rarely bigger than 20×20 nm^2 could be found on the surface. The low polymerization yield can be confidently assigned to the molecularity of the reaction in the second step, six functional groups being simultaneously involved in the Ullmann coupling of TDBPB. This issue was addressed in the next study where the number of halogens were reduced by a factor 2 compared to this study [95]. Similarly to DBPBA, the formation of hexagonal pores resulted from the combination of two consecutive steps, the building block being this time *p*-bromobenzeneboronic acid (BBBA). Due to its reduced molecular weight, the sublimation of BBBA onto Au(111) substrates could be realized at lower temperature than that required for the trimer of DBPBA (100 °C for BBBA vs. 190 °C for the trimer of DBPBA).

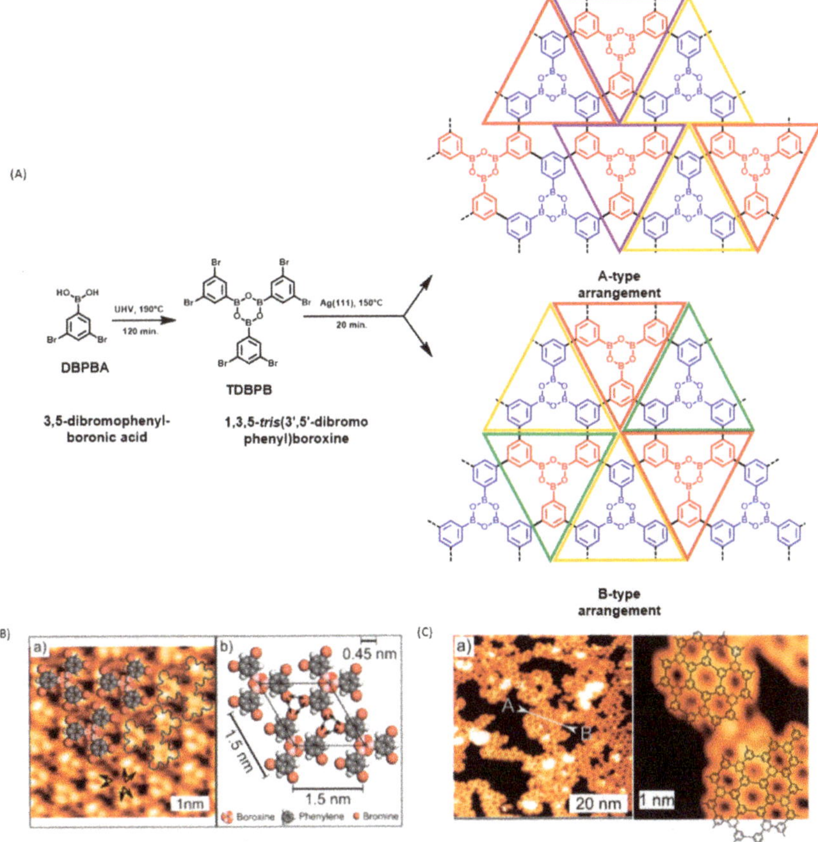

Figure 11. Reactions involved in the polymerization process of 3,5-dibromophenylboronic acid (DBPBA). (**A**) The two potential A- and B-type arrangements. (**B**) STM topograph of the supramolecular phase on Ag(111) obtained with 1,3,5-*tris*(3',5'-dibromophenyl)boroxine (TDBPB) and the tentative model of the monolayer. (**C**) STM topographs of the polymer phase at different magnifications. Reproduced from [94] with permission from The Royal Society of Chemistry.

Compared to DBPBA, the advantages offered by BBBA are manifold: (1) The low molecular weight of BBBA drastically limits the degradation and the probability of polymerization during sublimation. In fact, investigations revealed BBBA to be deposited intact on the Ag(111) substrates and to polymerize subsequent to its deposition on the Au(111) surface. The competition exiting between sublimation and polymerization for boronic acids is a well-known drawback of this family of compounds since only a minor fraction of boronic acids can be sublimed onto the surface, the rest polymerizing in the evaporator. (2) Limitation of the molecularity of the reaction at each step optimizes the probability of reaction and thus improves the reaction yields. Possibilities to create defects are considerably reduced. (3) Creation of hexagonal pores of large size becomes possible, the two-step growth enabling to avoid the evaporation of large precursors with important molecular weight. However, the detrimental role of the intermediate supramolecular phase was clearly evidenced.

Due to the formation of dense nanodomains of BBBA dimers stabilized by hydrogen bonds, the distance between trimers was not sufficient for the molecules to rearrange and diffuse onto the surface during annealing at 250 °C. If the formation of the polymer network was clearly evidenced, the reaction yield was limited to 88%, insufficient for the ring-closure of hexagons. Only 40% of pores covering the surface were detected (see Figure 12A).

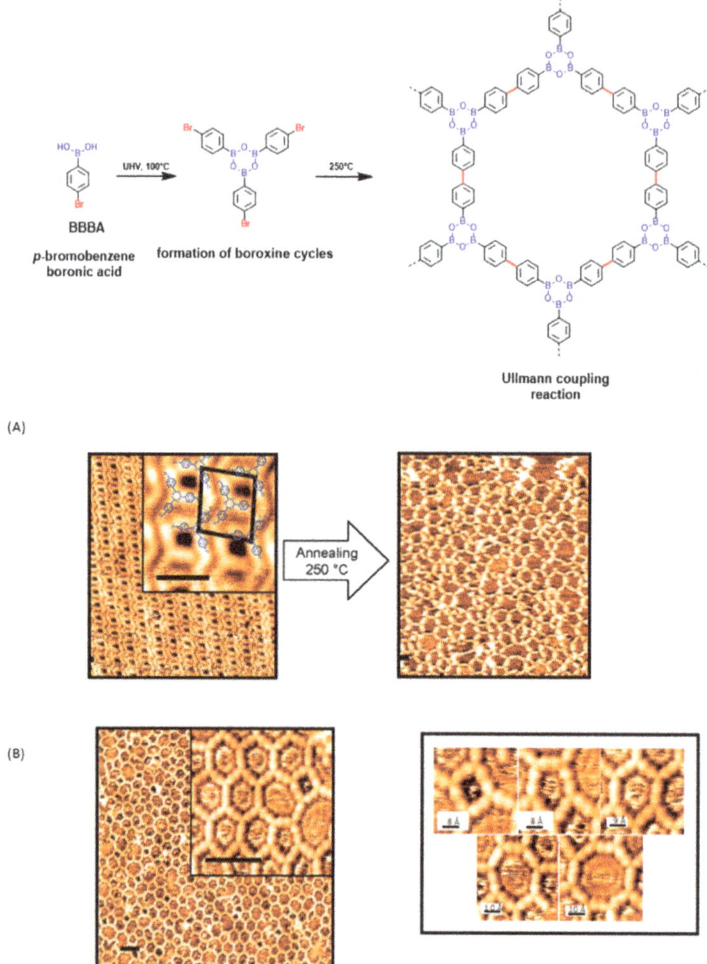

Figure 12. Polymerization reaction of *p*-bromobenzeneboronic acid (BBBA) on Au(111) substrates. (**A**) STM topographs of the supramolecular phase further annealed at 250 °C (10 nm × 14 nm, inset: scale bar = 2 nm). (**B**) STM topograph of the polymerization of BBBA in one step. Polygons of different shapes found on the Au(111) surface (left: scale bar = 4 nm). Adapted with permission from [95]. Copyright 2012 American Chemical Society.

In fact, investigations revealed the density of BBBA dimers in the supramolecular phase to be 1.5 BBBA/nm^2, higher than the density of BBBA in an ideal honeycomb-like structure (1.2 BBBA/nm^2). The material being denser in the supramolecular phase than in the polymer network, it clearly supports the formation of numerous defects during the conversion of the supramolecular assembly to the polymer. As a consequence of the high density of molecules on the surface and the insufficient distance between molecules to rearrange, polygons of different shapes could be detected, going from squares to octagons. Defects could not be repaired, even by a post-thermal treatment of the surface at 400 °C. The drawback issued from the close packing of the molecules in the supramolecular phase is an issue that has already been reported in the literature [29,96]. To address the density issue, BBBA was directly deposited on Au(111) substrates maintained at 250 °C. In these conditions, the two steps could be realized at low molecular density, favoring the diffusion and the rotation of the molecules on the surface and enabling in turn the molecules to be properly positioned for the formation of the boroxine cycles and the Ullmann couplings (see Figure 12B). As evidenced in the Figure 12B, an STM image with a scan area size larger than 55 × 55 nm^2 revealed the surface to be totally covered by the polymer, demonstrating that the polymerization can be carried out over a large area. A polymerization yield increasing up to 95% accompanied with the formation of 60% of pores were determined. Compared to the postannealing approach (first attempt), deposition of BBBA at high temperature favored the polymerization process. The total area covered by the polymer was doubled compared to the first attempts, even if a similar distribution of pore shapes was found in the two cases. Disorder in the final polymer was assigned to the size of the BBBA trimer, favoring the flexibility of the structure and the deviation from the regular hexagons. These conclusions are consistent with previous results reported in the literature for monomers of equivalent size [20,86]. Finally, examination of metal surfaces other than Au(111) revealed the choice of the surface to be essential to get well-separated temperature windows for the two steps. Thus, experiences carried out on Ag(111) revealed the two reactions to occur simultaneously, resulting in the formation of disordered structures on the surface.

2.3. Sequential Growth Based on the Boroxine Formation/Imine Combination

Capitalizing on the findings of the former studies, the possibility to combine two orthogonal reactions, namely, the formation of the boroxine cycles and imines in a single step was examined in 2016 on highly oriented pyrolytic graphite (HOPG) (see Figure 13) [97]. As the main difference from the previous works, the formation of sCOFs was examined in a multicomponent reaction with the amine and the aldehyde functional group attached to different precursors. Influence of the substitution pattern was also studied by differing the position of the aldehyde functional group.

As main advantages of these selected reactions, even if the two reaction pathways are activated simultaneously, the boroxine and imine formations are reversible reactions based on the release of water molecules which is the same side-product for the two reactions. In this context, self-healing opportunities during the polymerization process are greatly improved by the presence of numerous water molecules. With reference to "chemically released diffusion", self-healing typically refers to this concept, the possibility to repair defects resulting from mass transfer between polymer chains under growth. Considering that water molecules are exchanged between macromolecules under growth enabling to cleave and reform specific bonds, the term "chemically-assisted healing" would be more accurate in the present case.

To form the 2D structures, two bifunctional precursors were examined, namely 4-formylphenylboronic acid (4FPBA) and 3-formylphenylboronic acid (3FPBA) that were both opposed to 1,3,5-*tris* (4-aminophenyl)benzene (TAPB). Interestingly, two different chiral phases were obtained upon reaction of 3FPBA with TAPB and these results are among the first examples of chiral sCOF reported in the literature (see Figure 13). To perform these different two-component reactions, TAPB was first drop-casted onto the HOPG substrate whereas the second partner (4FPBA or 3FPBA) was placed in a reactor as powder in the presence of copper sulfate pentahydrate as the equilibrium control agent [92].

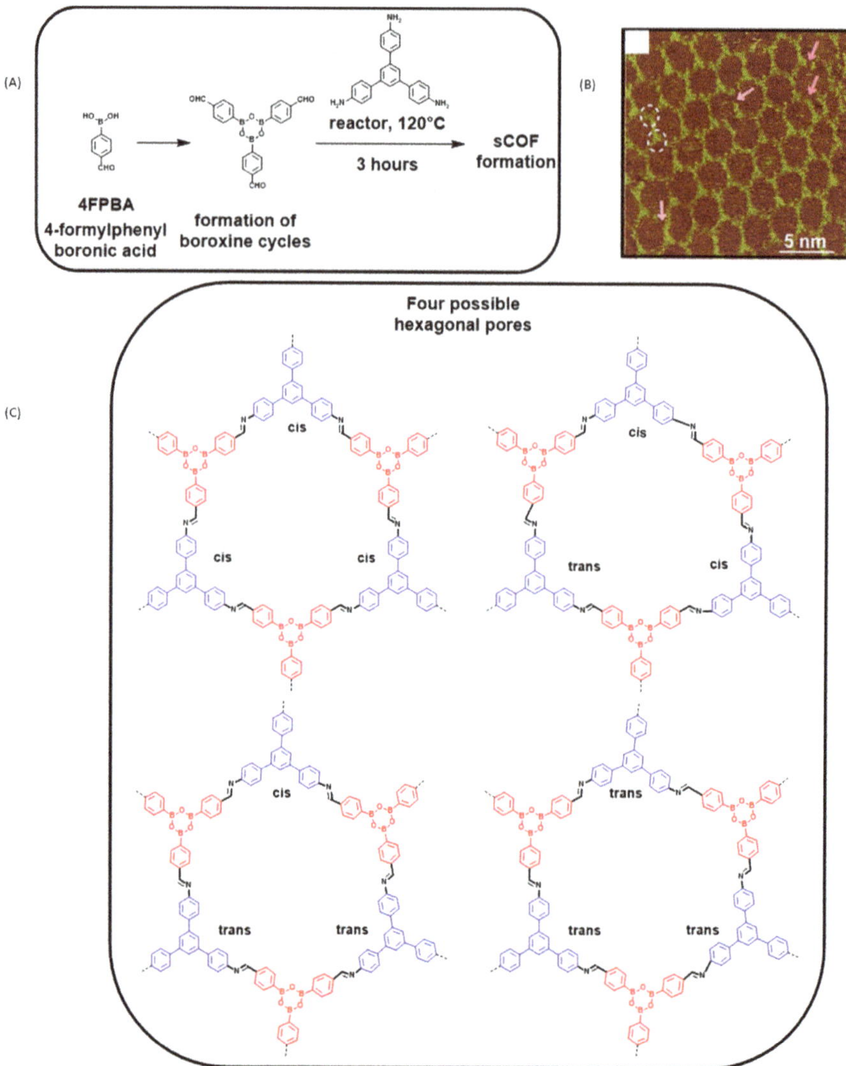

Figure 13. Polymerization reaction of 4-formylphenylboronic acid (4FPBA) with 1,3,5-*tris*(4-aminophenyl)benzene (TAPB). (**A**) The two reaction steps. (**B**) STM topograph of the polymer network. (**C**) the different possible conformations of the imine groups and the resulting hexagons. Reproduced from [97] with permission from The Royal Society of Chemistry.

When 4FPBA was used as the precursor, three hours at 120 °C in a sealed reactor were required to obtain a polymerization. Typically, nanodomains of 80×80 nm^2 were detected, the growth of these domains being orientated along the lattice of the HOPG surface (see Figure 13B). Due to the possible cis/trans isomerization of the imine functions, an inherent disorder exists and the presence of distorted hexagons on the surface is clearly detected (see Figure 13C). As the main drawback of employing functional groups with a nonpermanent orientation, presence of unclosed hexagons was observed on the HOPG substrate due to the inappropriate orientation of numerous imine functional groups.

A more complicated behavior was evidenced for 3FPBA. Indeed, due to the asymmetrical substitution of 3FPBA, a chirality can appear due to the formation of two types of 3FPBA trimers that can rotate clockwise (CW-3FPBA) or counterclockwise (CCW-3FPBA) (see Figure 14A). By polymerization with TAPB, the coexistence of two chiral nanodomains with a long-range order was found on the surface) (see Figure 14B). As specificity, CW-3FPBA furnished a CCW-sCOF, whereas the opposite situation was found for CCW-3FPBA (CW-sCOF). For the two chiral sCOFs, a growth orientation governed by the interaction of TAPB with the substrate was found, imposing the epitaxial conditions of growth.

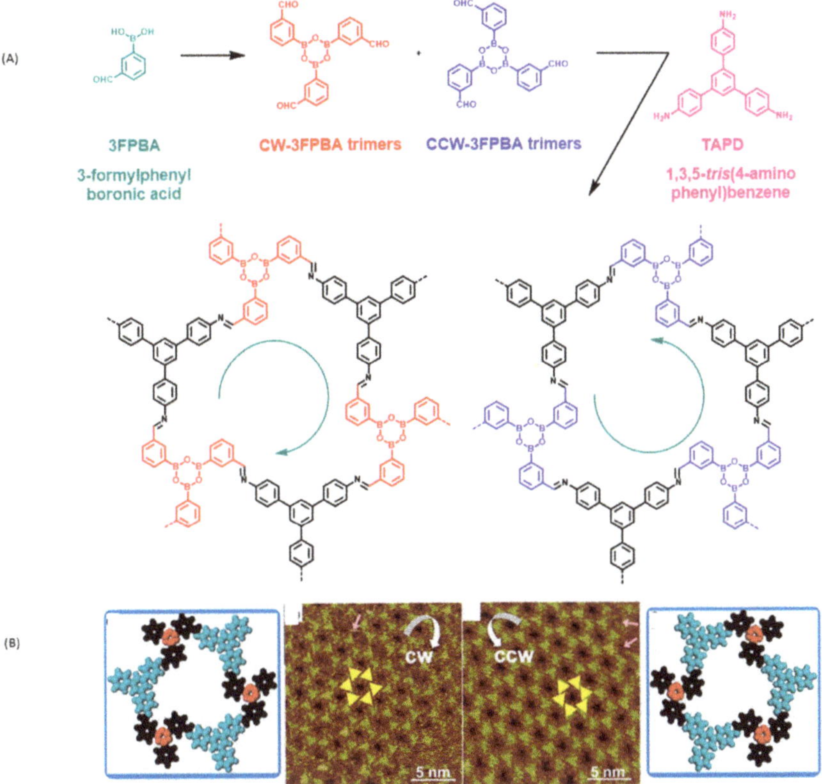

Figure 14. (**A**) Reaction involved in the polymerization process. Clockwise (CW) and counter-clockwise (CCW) rotating 3FPA trimers. (**B**) STM topographs of the sCOF obtained with 3-formylphenylboronic acid (3FPBA) and 1,3,5-*tris*(4-aminophenyl)benzene (TAPB), and the demonstration of the presence of two chiral phases. Reproduced from [97] with permission from The Royal Society of Chemistry.

Evidence of imines formation was furnished by X-ray photoelectron spectroscopy (XPS) analyses. A peak at 398.5 eV characteristic for the nitrogen of an imine group was detected in the XPS spectrum. Conversely, a peak at 399.8 eV was found for the precursor TAPB, demonstrating the chemical modification of the NH_2 group. The crucial role of copper sulfate pentahydrate was clearly demonstrated in control experiments. Notably, a higher structural disorder accompanied by incomplete reactions was found for the polymerization of 4FPBA with TAPB in the absence of this salt (see Figure 15A).

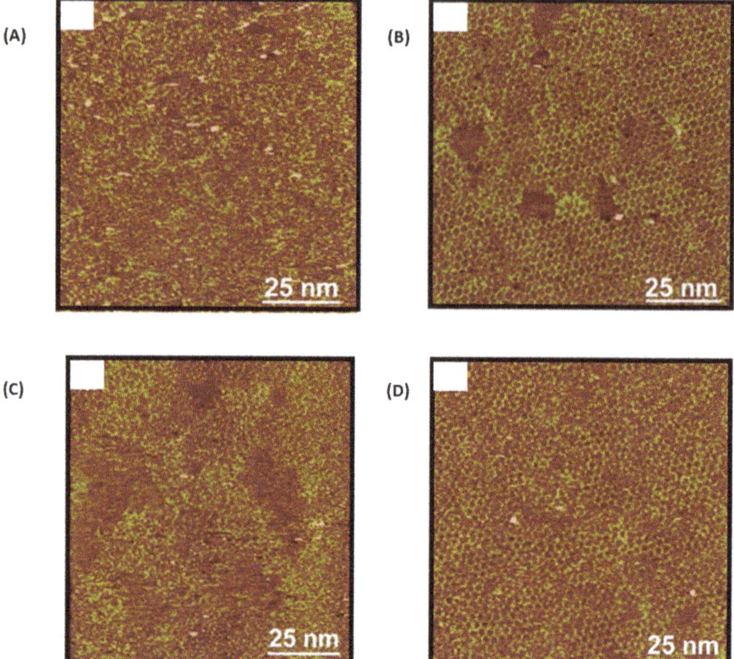

Figure 15. (**A**) STM topograph of the polymerization of 4-formylphenylboronic acid (4FPBA) with 1,3,5-*tris*(4-aminophenyl)benzene (TAPB) without copper sulfate pentahydrate. (**B**) STM topograph of the polymerization of 4FPBA with TAPB for a reaction carried out at 120 °C for 3 h. (**C**) Stability of the 2D polymer formed with 4FPBA and TAPB after heating at 180 °C for 1 h. (**D**) Stability of the polymer formed from 4FPBA and TAPB after 20 days of storage. Reproduced from [97] with permission from The Royal Society of Chemistry.

In fact, this hydrated salt is capable in a sealed reactor to release water and favor the reversibility of the imine and boroxine formation. Examination of the kinetic of polymerization of 4FPBA with TAPB also revealed the boroxine cycle to be formed at the same timescale than the imine functional groups, and that a high coverage of the HOPG surface can only be obtained after four hours of reaction (see Figure 15B). Finally, stability of the sCOFs were examined and several conclusions could be determined. Thus, sCOFs formed with 4FPBA were stable after annealing of the surface for one hour at 180 °C, demonstrating the robustness of the structure (see Figure 15C). Second, after 20 days of storage under ambient conditions, sCOFs formed with 4FPBA could be still observed on the substrates what is remarkable considering that the boroxine cycle is easy to hydrolyze (see Figure 15D). This stability is higher than that previously reported for sCOF only formed of boroxine cycles [92]. However, examination of the chemical stability of these structures revealed that exposition of sCOF to acidic (pH < 3) or basic (pH > 11) conditions rapidly furnished chaotic structures.

2.4. Hierarchical Growth Based on the Coordination Polymer/Phthalocyanine Formation

A longstanding challenge toward the development of advanced functional materials relies in the possibility to create fully conjugated covalent networks. An interesting approach consisted of creating in situ the phthalocyanine macrocycles in successive steps involving first the formation of a supramolecular phase further converted to a covalent polymer by thermal annealing [98]. Phthalocyanines are a family of compounds extensively used in organic electronics by the possibility to finely tune their electronic properties by a careful selection of the metal center. These metal complexes are also characterized by a remarkable thermal and chemical stability, making these structures appealing candidates for numerous applications. Parallel to this, numerous examples of on-surface chemistry based on the phthalocyanine motif have been reported in the literature [99]. Here, and contrarily to the former study reported in 2011 where a polymer of Fe-phthalocyanine was formed in one step on Au(111) [100], in this subsequent study carried out by the same authors, the different intermediate steps could be isolated and characterized (see Figure 16). As the main difference, Mn instead of Fe was used as the metal, Ag(111) substrate was used instead of Au(111), while 1,2,4,5-tetracyanobenzene (TCNB) was used as the same elemental building block. In this work, formation of the phthalocyanine polymer could be decomposed in three steps, the first one consisting in the sublimation of both the metal and the precursor on Ag(111) held at room temperature, generating a two-dimensional coordination polymer with a square symmetry (see Figure 17A). In this first 2D network, Mn and TCNB could be found in a 1:1 ratio. Upon annealing at ca 377 °C, Mn-based octacyanophthalocyanines arranged in a closely packed supramolecular structure self-assembled by hydrogen bonds and aligned along the dense directions of the substrate were found (see Figure 17B).

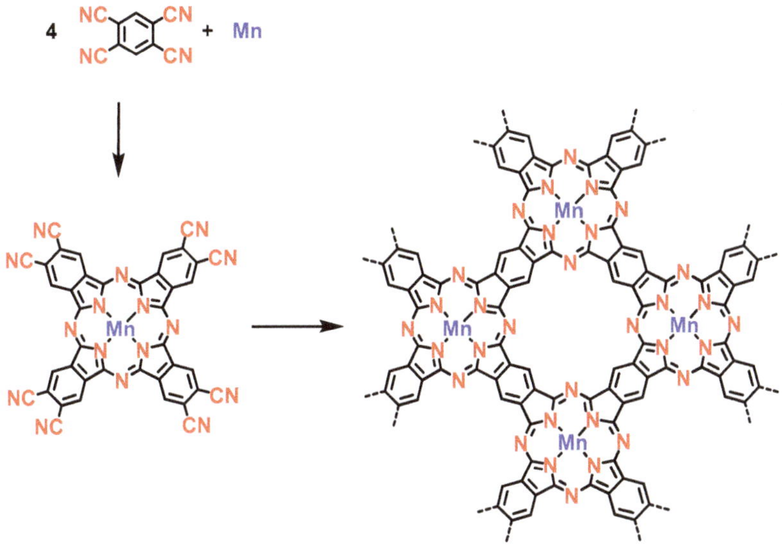

Figure 16. Sequential growth of the phthalocyanine-based polymer.

Coexisting with this first phase, a second phase, where Mn-based octoacyanophthalocyanines are linked by mean of Mn atoms, could also be found. Considering that Mn-based octoacyano-phthalocyanines can be formed in a 4:1 TCNB:Mn ratio, the presence of this second phase resulting from metal–ligand interactions between the electron-donating nitrile groups and the electron-deficient metal atoms could be attended. It also supports the formation in the final stage (annealing at 342 °C) of small 5-nm-large domains composed of polymeric phthalocyanines, where the presence of free Mn atoms is necessary to form the additional phthalocyanines and form the fully conjugated polymer (see Figure 17C).

In this 2D polymeric phase, magnetic atoms regularly spaced and introduced in a fully π-conjugated system could be obtained. This work opens the way towards the bottom-up elaboration of devices for data storage.

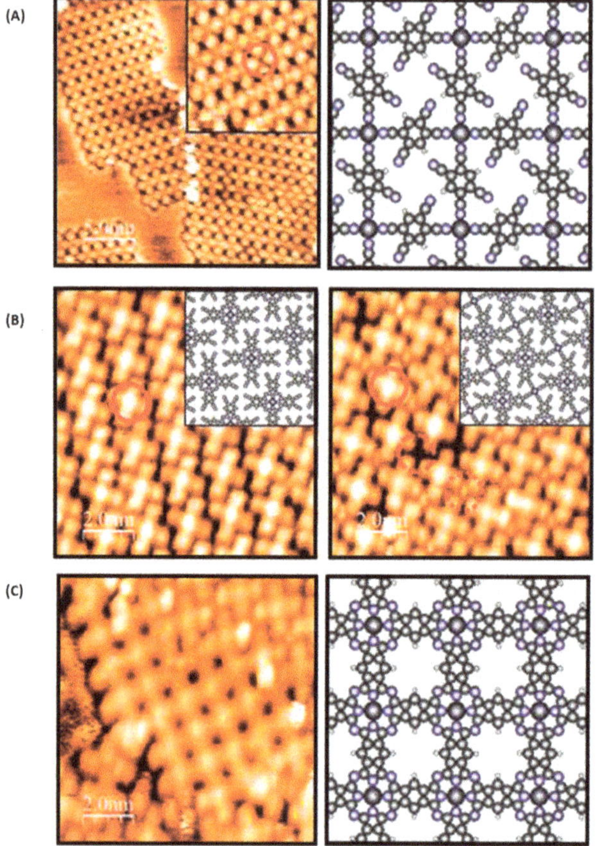

Figure 17. The different steps involved in the formation of phthalocyanines-based 2D π-conjugated polymers. (**A**) STM images of the supramolecular phase resulting from the deposition of 1,2,4,5-tetracyanobenzene (TCNB) and Mn atoms on a Ag(111) surface at room temperature with a DFT model. (**B**) STM images of the two phases existing after annealing at 277 °C. Left: the hydrogen-bonded supramolecular phase. Right: the metal–organic coordination 2D polymers. (**C**) Polymeric phase obtained in the final step upon annealing at 342 °C. Reproduced from [98] with permission from The Royal Society of Chemistry.

3. Coupling Modes Used for the Design of 1D Macromolecular Organic Structures

3.1. Sequential Growth Based on the Ullmann Coupling/Aromatization Combination

In all the aforementioned reaction combinations, 2D covalent networks have been obtained by sequential or hierarchical growth of nanostructures. However, a series of combinations has also been specifically developed for the design of 1D macromolecular organic structures. At present, it has to be noticed that these combinations have not been used for the design of 2D structures yet, even if from a technical point of view, these combinations could be easily transposed to 2D structures. In this field, graphene, by its unique properties, has driven a great deal of interest and numerous works have been devoted to design graphene-like structures. To produce regular and extended structures, the Ullmann coupling followed by an aromatization reaction is a promising approach (see Figure 18).

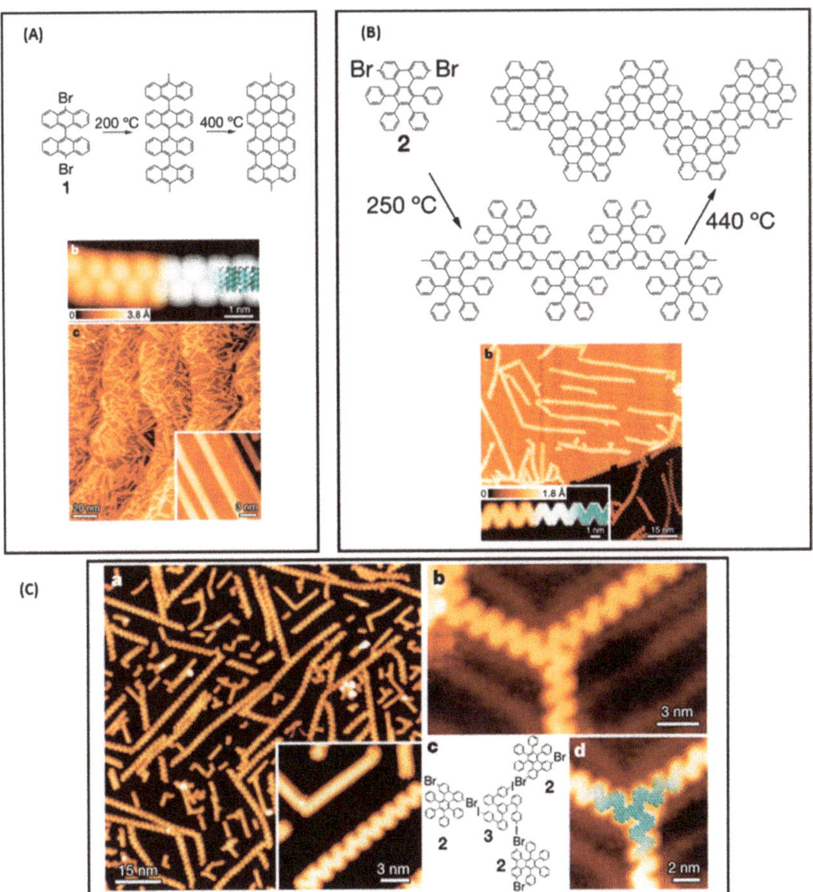

Figure 18. (**A**) Reaction mechanism and STM images of the formation of nanoribbons with 10,10′-dibromo-9,9′-bianthracene. (**B**) Reaction mechanism and STM images of the formation of zigzag nanoribbons with 6,11-dibromo-1,2,3,4-tetraphenyltriphenylene. (**C**) STM images of the copolymerization reaction. All experiments have been carried out on Au(111) surfaces. Reprinted by permission from Macmillan Publishers Ltd: Nature from [101], copyright 2010.

In this field, the first on-surface synthesis was reported in 2010 [101]. Graphene nanoribbons (GNRs) were obtained by first depositing 10,10′-dibromo-9,9′-bianthracene on Au(111) surfaces at 200 °C, enabling to interlink the biradicals resulting from the dehalogenation reaction, the planarization of the structure and the formation of a linear polymer. At 400 °C, a cyclodehydrogenation reaction converted the polymer to a fully conjugated and linear structure (see Figure 18A). Versatility of the approach was demonstrated by the design of several types of nanoribbons, such as chevron-type nanoribbons resulting from the polymerization and aromatization reaction of 6,11-dibromo-1,2,3,4-tetraphenyltriphenylene (see Figure 18B). In this last case, steric hindrance generated by the tetraphenylene groups enforces these bulky substituents to stand on each side of the polymer main axis and a zigzag structure was obtained. The possibility of codepositing two differing precursors (6,11-dibromo-1,2,3,4-tetraphenyltriphenylene and 1,3,5-*tris*(4″-iodo-2′-biphenyl)benzene) was also examined, demonstrating the possibility to design chemically modified graphene-like structures (see Figure 18C). Recently, the different experiments carried out by Cai et al. on 10,10′-

dibromo-9,9′-bianthracene and 10,10′-dibromo-9,9′-bianthracene were reproduced using another deposition process for the first step. In this last case, the direct contact transfer (DCT) consisting in using a stamp with the corresponding molecule at its surface was used [102]. Using this strategy, the precursor could be deposited on the surface without taking recourse to sublimation. This finding constitutes a major advancement, considering that the molecular weight often constitutes a drastic limitation. Indeed, increase of the molecular size results in higher sublimation temperatures that can degrade the molecule. Over the years, a wide range of molecular tectons giving access to GNRs have been reported in the literature and these different structures are depicted in the Figure 19 [103–117].

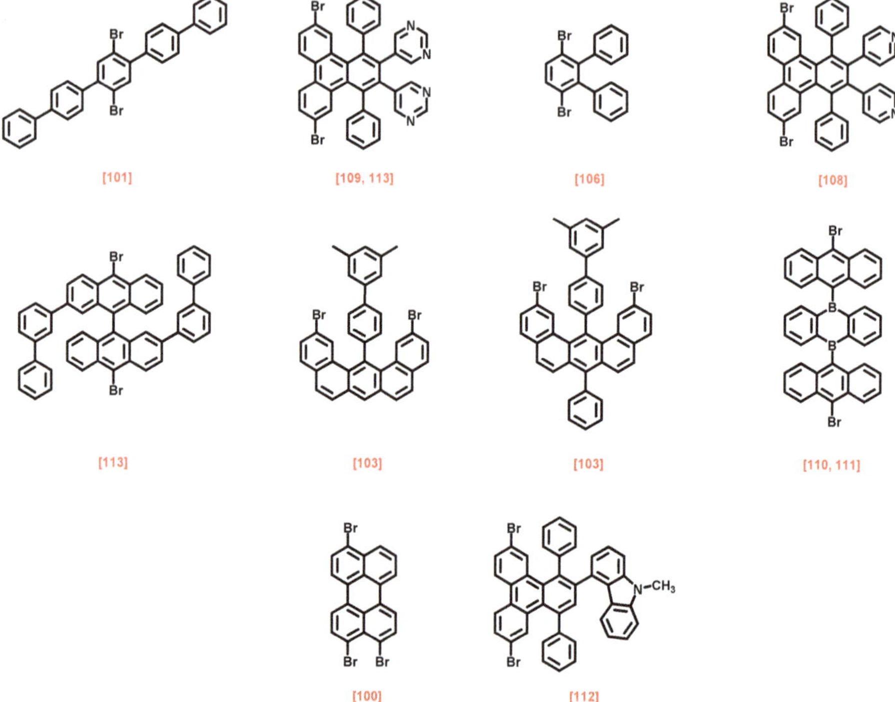

Figure 19. Chemical structures of precursors used to elaborate graphene nanoribbons (GNRs) and the corresponding reference in brackets.

The beauty of these approaches relies in the fact that nitrogen-doped GNRs [111,116], ultranarrow GNRs [103,118], chevron-type GNRs [102], or zigzag GNRs [106,117,119] with chemically modified edges could be elaborated with these precursors. Among the most interesting findings and by properly choosing the molecular precursors, GNRs could notably be used as elemental building blocks to trigger the design of more extended structures. This goal was achieved by substituting the precursors with nitrogen atoms [112]. Ordered nanomaterials could be obtained by an atomically precise bottom-up synthesis based on the creation of hydrogen bonds and van der Waals interactions between GNRs (see Figure 20A). Growth of structures in the third dimension could be even obtained by π–π stacking of GNRs sheets. Modification of the electronic structures of GNRs is another long-standing challenge and, in this field, a remarkable example has been reported in 2016 [115].

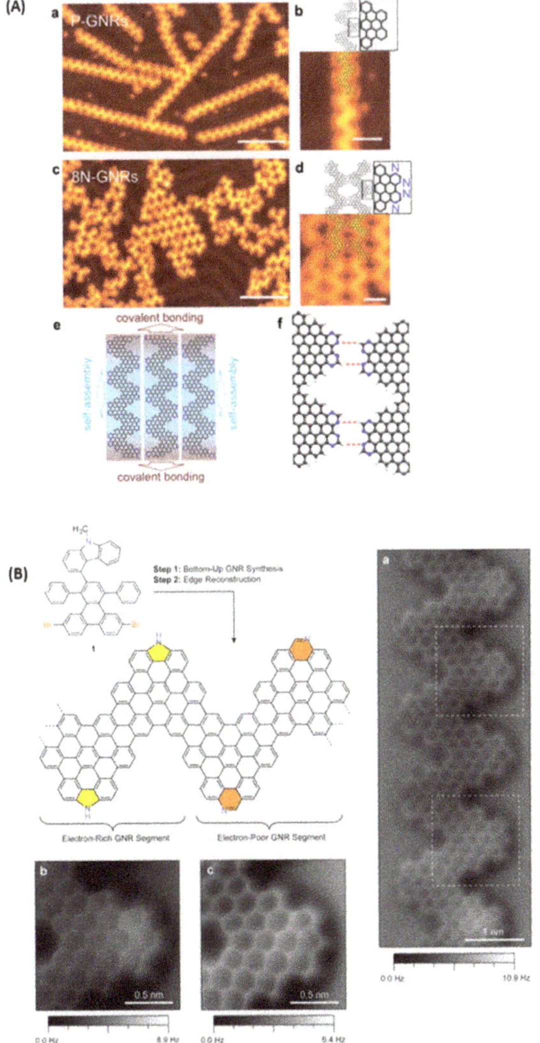

Figure 20. (**Aa–f**) Bottom-up synthesis of GNRs containing nitrogen atoms with the possibility to create lateral interactions between GNRs (a: scale bar: 10 nm, b: scale bar: 2 nm, c: scale bar: 10 nm, d: scale bar: 2 nm). Adapted with permission from Vo et al. [112]. Copyright 2015 American Chemical Society. (**B**) Schematic representation of GNRs comprising both carbazole and phenanthridine units. (**Ba–c**) High-resolution AFM images of GNRs with the demonstration of the presence of phenanthridine (**Bb**) and carbazole (**Bc**) units in GNRs. Reproduced from [113] with permission from The Royal Society of Chemistry.

By a thermally activated ring expansion/dehydrogenation reaction, the electron-rich carbazole could be converted into the electron-deficient phenanthridine [120]. In this work, authors succeeded to partially convert the carbazole groups into phenanthridine, opening the way towards the fine tuning of the materials bandgaps by the presence of electron-rich and electron-poor groups onto the same structures (see Figure 20B).

Choice of the surface as well as a careful selection of the shape of the molecular tectons (planar or nonplanar) can impact the geometry of the final GNRs and greatly help in designing GNRs with precise structures [121]. These opportunities were demonstrated with an extensive study carried out on two molecular building blocks, i.e., 10,10′-dibromo-9,9′-bianthracene (DBBA). The debate concerning the molecular arrangement of DBBA on Cu(111) is a long-standing issue [105,107,108,122,123] and by using noncontact atomic force microscopy (nc-AFM), the controversy could be definitely solved. In the specific case of DBBA and despites the presence of bromine atoms, the Ullmann reaction proved to be ineffective on Cu(111) substrates to couple the molecular tectons and GNRs resulting from another coupling mode could be detected on the surface. Interestingly, the same outcome than the one obtained with DBBA could be produced with the unsubstituted 9,9′-bianthracene (BA) or 10,10′-dichloro-9,9′-bianthracene on Cu(111) substrates (see Figure 21Aa–Ad).

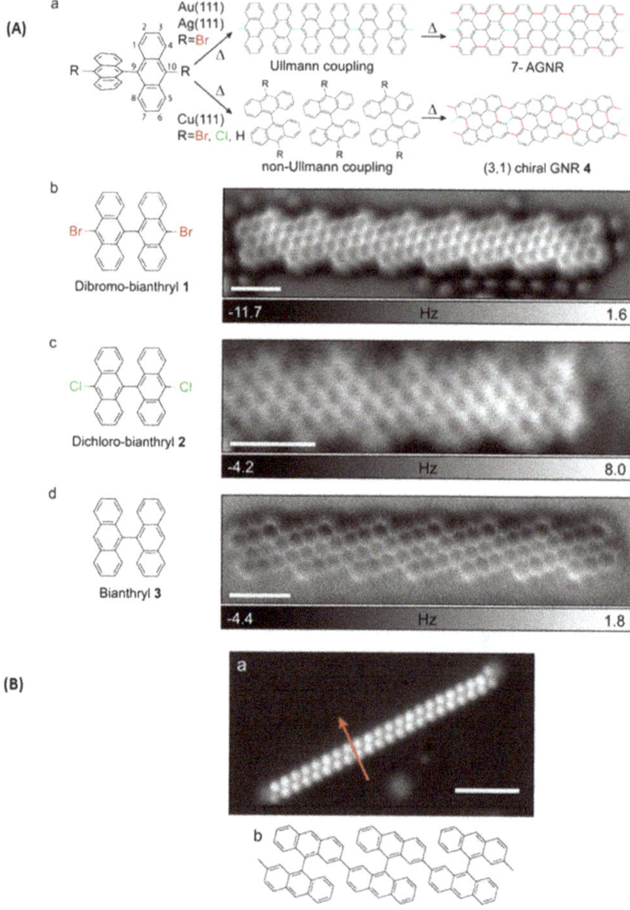

Figure 21. (**A**) GNRs obtained with different bianthracene (BA) precursors. (**B**) Connection of bianthracene precursors at the C2 and C2′ positions (scale bar: 1 nm for all noncontact AFM images). Adapted with permission from Schulz et al. [121]. Copyright 2017 American Chemical Society.

Conversely, upon deposition of DBBA on Ag(111) or Au(111), the conventional Ullmann coupling reaction occurred, providing GNRs with armchair edges. By combining STM and nc-AFM at low temperature, the mechanism could be elucidated. Authors demonstrated that upon cleavage of the carbon–halogen bond, the biradicals of DBBA were stabilized by the strong interactions existing with the Cu(111) surface and/or adatoms, so that a severe reduction of the energy barrier for the carbon–hydrogen bond scission at the C2 and C2′ positions could be obtained. Parallel to this, the twisted structures of the different BA derivatives enforced a specific arrangement of the molecular tectons on the surface, favorable to intermolecular interactions at the C2/C2′ positions (see Figure 21(Ba,Bb). As a result of this, and irrespective of the substitution pattern of BA, the same GNRs could be obtained with all monomers, the homocoupling being governed by the C–H scission and not by the recombination of radicals.

3.2. Sequential Growth Based on the Coordination Polymer/Glaser Coupling Combination

Still implying intermediate organometallic structures in multistep reactions, a surprising example has been reported in 2018 where the monomer, i.e., 2,5-diethynyl-1,4-*bis*(4-bromophenyl-ethynyl) benzene (2Br-DEBPB) could theoretically react according to the Ullmann and Glaser coupling mechanisms (see Figure 22A) [124].

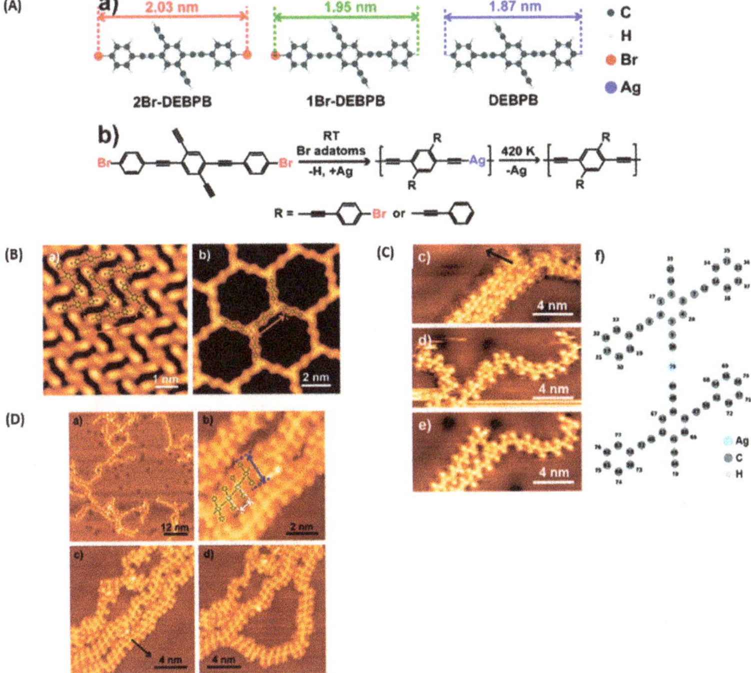

Figure 22. (**A**) Polymerization of 2,5-diethynyl-1,4-*bis*(4-bromophenyl-ethynyl)benzene (2Br-DEBPB) on Ag(111) substrates and chemical structures of 1-((4-bromophenyl)ethynyl)-2,5-diethynyl-4-(phenylethynyl)benzene (1Br-DEBPB) and ((2,5-diethynyl-1,4-phenylene)bis(ethyne-2,1-diyl))dibenzene (DEBPB). (**B**) STM images of the supramolecular phases obtained by annealing the surface at RT temperature for 30 min. (**C**) STM images of the coordination polymer obtained by annealing the surface at RT temperature for 10.5 hours. 4) Covalent polymer chains obtained by annealing the Ag(111) surface at 410K. Reproduced from [124] with permission from The Royal Society of Chemistry.

However, only one type of coupling was observed on Ag(111) surfaces. To get this discrimination between the two possible types of coupling, the experiments were carried out at low temperature. Thus, coexistence of two supramolecular phases consisting in a chevron-type structure and a honeycomb arrangement could be found on a Ag(111) surface after a thermal evaporation of 2Br-DEBPB on a surface held at $-123\,^{\circ}$C (see Figure 22B). After annealing this surface for 1.5 h at room temperature and subsequent cooling of the surface at low temperature (see Figure 22C), a newly ordered structure could be found on Ag (111) surfaces, consisting of parallel rows of molecules where intact molecules and monobrominated 1-((4-bromophenyl)ethynyl)-2,5-diethynyl-4-(phenylethynyl)benzene (1Br-DEBPB) (2% of the total molecules) were coassembled in a densely packed structure. Presence of a large number of bromine adatoms (higher than the 2% resulting from the debromination reaction) on the surface was assigned to the desorption of the fully debrominated molecules. This fact was confirmed by annealing the supramolecular phase for 30 min instead of 1.5 h at room temperature. Twenty-one percent of fully reduced molecules could be found on the surface. However, it has to be noticed that the temperature was insufficient to initiate an Ullmann coupling. Molecular chains corresponding to the formation of coordination polymers could only be obtained by annealing the surface for 10 hours at RT. In these conditions, an activation of the C–H bonds of terminal alkynes could be obtained, resulting in the formation of polymer chains. Finally, conversion of the coordination polymers to covalently bonded polymers was achieved at the relatively low temperature of $137\,^{\circ}$C (See Figure 22D). By the abundant presence of bromine adatoms on the Ag(111) surface, interactions between the electron-rich bromines and the electron-deficient hydrogens of terminal alkynes could occur, reducing the BDE, and weakening the C–H bond. As a result of this, Glaser coupling, assisted by both the surface and the presence of halogen adatoms, could occur.

3.3. Sequential Growth Based on the Glaser Coupling/Dehydrogenative Coupling Combination

Convincing results concerning the benefits of a sequential procedure were also obtained by combining a Glaser coupling with a dehydrogenative polymerization [125]. Use of carboxylic acid derivatives for on-surface syntheses of supramolecular networks is not new and this is notably due to the ability of the carboxylic functional groups to form hydrogen bonds or to interact with the metal substrates, giving rise to metal–ligand coordination networks [126,127]. The carboxylic functional groups can also be deprotonated by the metal surface, drastically modifying the structure of the final network. A clear evidence of this influence was provided with trimesic acid deposited onto a Cu(100) surface [128]. Deprotonation of the carboxylic functions modified the adsorption geometry of the network by inducing a rotation of 90° relative to the substrate so that the deprotonated form of trimesic acid stands upright and perpendicular to the surface. With aim at generating covalent bonds, only few articles have been reported in the literature and the decarboxylation of 2,6-naphthalenedicarboxylic acid to form carbon–carbon bonds can be cited as the unique example [129]. Parallel to this, on-surface Glaser coupling of terminal alkynes was reported prior to this work but examples in the literature are scarce [42]. Orthogonality of these two reactions is clear so that their combination was examined for the polymerization of 6-ethynyl-2-naphthoic acid (ENA). Interestingly, influence of the surface topography as well as the density of molecules on surface were determined as controlling the reaction outcome. Upon deposition of ENA on Au(111) substrates at low surface coverage and upon annealing of the surface at $124\,^{\circ}$C, a 2D network formed, resulting from the dimerization of ENA, i.e., 6,6′-(buta-1,3-diyne-1,4-diyl)*bis*(2-naphthoic acid) (DBNA) and the formation of a metal–carboxylate coordination network (see Figure 23). Differing from this first arrangement, the same experiments carried out at high surface coverage resulted in the coexistence of two distinct phases on the surface (See Figure 24A). The first one corresponds to an intermediate state of the Glaser dimerization where two ENA molecules are arranged in a linear fashion with Au adatoms ensuring the connection with two ENA units. In the second phase and contrarily to what was expected, the product resulting from the Glaser coupling, i.e., DBNA was detected on the surface, but coexisting with an unexpected compound resulting from an oxidative dehydrogenation. In fact, disordered polymer chains could be detected in this second phase, resulting from the formation of bisacylperoxides (see Figure 24B).

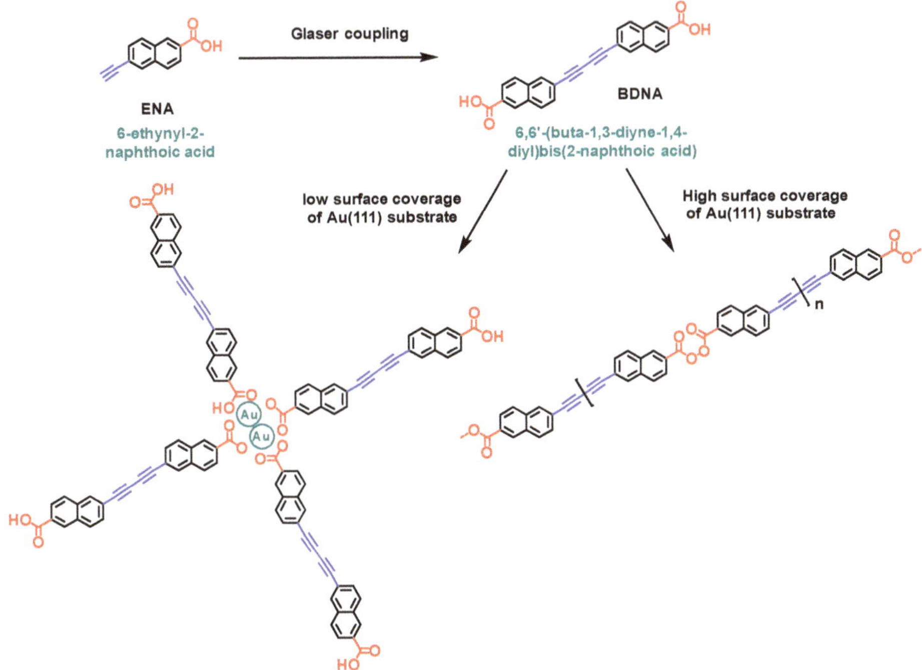

Figure 23. Polymerization of 6-ethynyl-2-naphthoic acid (ENA) by a domino reaction based on the Glaser coupling/dehydrogenative coupling combination. Depending of the density of molecules onto the Au(111) substrates, different structures are obtained.

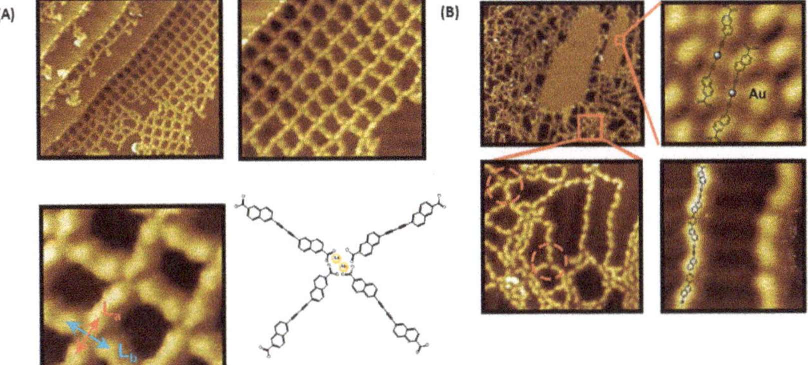

Figure 24. Polymerization of 6-ethynyl-2-naphthoic acid (ENA) by a domino reaction based on the Glaser coupling/dehydrogenative coupling combination. (**A**) STM topographs obtained at low density of molecules after Glaser coupling and formation of the Au-carboxylate complex on Au(111) substrates (left: 42 nm × 38 nm, right: 21 nm × 17 nm, bottom: 5.9 nm × 5.0 nm). (**B**) STM topographs obtained at high density of molecules on Au(111) substrates (top left: 72 nm × 88 nm, top right: 2.6 nm × 1.2 nm, bottom left: 17 nm × 12 nm, bottom right: 5 nm × 4 nm). Reprinted with permission from Held et al. [125]. Copyright © 2016 John Wiley & Sons, Inc.

Interestingly, the starting point of these different polymer chains was still a gold–carboxylate complex. As a result of this, the direction of the chains growth is thus predetermined by the geometry of the Au-complex and the number of "ligands" around the Au center. To verify the role of the substrate in the growth direction, similar experiments were carried out onto Au(100) substrates with a structured surface and these latter revealed the Glaser coupling as well as the dehydrogenative coupling to produce chains aligned along with the channels direction, without the formation of ramifications. Polymers extending over 100 nm could be thus prepared, with a thermal stability higher than 160 °C. In this work, the crucial role of the first step, i.e., the formation of the butadiyne groups, was determined as being primordial to activate the formation of bisacylperoxides. Indeed, despite numerous aromatic acids having been examined for on-surface synthesis prior to this work, formation of peroxides from carboxylic groups has never been reported. Finally, XPS experiments confirmed the formation of both Au–carboxylate complexes as well as bisacylperoxide groups.

4. Conclusions

In this review, a series of eight combinations of orthogonal reactions enabling the successful realization of 1D and 2D macromolecular organic structures on surface have been reported. The development of efficient chemistries in a vacuum environment is an important challenge conditioning the future of numerous applications. For these reasons, it is of high importance to continue to acquire knowledge and know-how on the growth of macromolecular organic structures in vacuum environment. Based on the different works devoted to the hierarchical or sequential construction of 1D and 2D structures, the possibility to elaborate extended covalent organic structures on surface was brilliantly demonstrated with STM images of scan area sizes larger than 60×60 nm^2. Comparison between the macromolecular structures obtained by sequential or hierarchical growth on surface and those obtained by the classical one-step approach would be of crucial interest for the future development of this approach and the optimization of the reaction conditions. However, the comparison between the two approaches remains extremely difficult with regards to the multiparametric character (metal, crystallographic plane, formation or not of an intermediate supramolecular phase, density of molecules on-surface, etc.) of the delicate growth. This is notably the reason why the comparison between the two models is not established in the different studies. With regards to the large number of reactions existing in organic chemistry and the infinite of combinations, there is still room for improvement and numerous unexplored combinations will be reported in the future.

Author Contributions: Writing—Original Draft Preparation, C.P. and F.D.; Writing—Review & Editing, C.P. and F.D.

Funding: This research was funded by Aix Marseille University and the Centre National de la Recherche Scientifique (CNRS).

Conflicts of Interest: The authors declare no conflicts of interest.

References

1. Zhao, F.; Liu, H.; Mathe, S.D.R.; Dong, A.; Zhang, J. Covalent organic frameworks: From materials design to biomedical application. *Nanomaterials* **2018**, *8*, 15. [CrossRef] [PubMed]
2. Lohse, M.S.; Bein, T. Covalent organic frameworks: Structures, synthesis, and applications. *Adv. Funct. Mater.* **2018**, *28*, 1705553. [CrossRef]
3. Lackinger, M.; Griessl, S.; Markert, T.; Jamitzky, F.; Heckl, W.M. Self-assembly of benzene-dicarboxylic acid isomers at the liquid solid interface: Steric aspects of hydrogen bonding. *J. Phys. Chem. B* **2004**, *108*, 13652–13655. [CrossRef]
4. Griessl, S.; Lackinger, M.; Edelwirth, M.; Hietschold, M.; Heckl, W.M. Self-assembled two-dimensional molecular host-guest architectures from trimesic acid. *Single Mol.* **2002**, *3*, 25–31. [CrossRef]
5. Kühne, D.; Klappenberger, F.; Decker, R.; Schlickum, U.; Brune, H.; Klyatskaya, D.; Ruben, M.; Barth, J.V. High-quality 2D metal-organic coordination network providing giant cavities within mesoscale domains. *J. Am. Chem. Soc.* **2009**, *131*, 3881–3883. [CrossRef] [PubMed]

6. Langner, A.; Tait, S.L.; Lin, N.; Rajadurai, C.; Ruben, M.; Kern, K. Self-recognition and self-selection in multicomponent supramolecular coordination networks on surfaces. *Proc. Natl. Acad. Sci. USA* **2007**, *104*, 17927–17930. [CrossRef] [PubMed]

7. Barth, J.V. Molecular architectonic on metal surfaces. *Annu. Rev. Phys. Chem.* **2007**, *58*, 375–407. [CrossRef]

8. De Feyter, S.; De Schryver, F.C. Two-dimensional supramolecular self-assembly probed by scanning tunneling microscopy. *Chem. Soc. Rev.* **2003**, *32*, 139–150. [CrossRef]

9. Barth, J.V. Fresh perspectives for surface coordination chemistry. *Surf. Sci.* **2009**, *603*, 1533–1541. [CrossRef]

10. Koudia, M.; Nardi, E.; Siri, O.; Abel, M. On-surface synthesis of covalent coordination polymers on micrometer scale. *Nano Res.* **2017**, *10*, 933–940. [CrossRef]

11. Nakanishi, T. Supramolecular soft and hard materials based on self-assembly algorithms of alkyl-conjugated fullerenes. *Chem. Commun.* **2010**, *46*, 3425–3436. [CrossRef] [PubMed]

12. Dong, R.; Pfeffermann, M.; Liang, H.; Zheng, Z.; Zhu, X.; Zhang, J.; Feng, X. Large-area, free-standing, two-dimensional supramolecular polymer single-layer sheets for highly efficient electrocatalytic hydrogen evolution. *Angew. Chem. Int. Ed.* **2015**, *54*, 12058–12063. [CrossRef] [PubMed]

13. Ko, M.; Mendecki, L.; Mirica, K.A. Conductive two-dimensional metal–organic frameworks as multifunctional materials. *Chem. Commun.* **2018**, *54*, 7873–7891. [CrossRef] [PubMed]

14. Gourdon, A. On-surface covalent coupling in ultrahigh vacuum. *Angew. Chem. Int. Ed.* **2008**, *47*, 6950–6953. [CrossRef] [PubMed]

15. Perepichka, D.F.; Rosei, F. Extending polymer conjugation into the second dimension. *Science* **2009**, *323*, 216–217. [CrossRef] [PubMed]

16. Mendez, J.; Lopez, M.F.; Martin-Gago, J.A. On-surface synthesis of cyclic organic molecules. *Chem. Soc. Rev.* **2011**, *40*, 4578–4590. [CrossRef] [PubMed]

17. Franc, G.; Gourdon, A. Covalent networks through on-surface chemistry in ultrahigh vacuum: State-of-the-art and recent developments. *Phys. Chem. Chem. Phys.* **2011**, *13*, 14283–14292. [CrossRef] [PubMed]

18. Grill, L.; Dyer, M.; Lafferentz, L.; Persson, M.; Peters, M.V.; Hecht, S. Nano-architectures by covalent assembly of molecular building blocks. *Nat. Nanotechnol.* **2007**, *2*, 687–691. [CrossRef] [PubMed]

19. Bieri, M.; Treier, M.; Cai, J.; Aït-Mansour, K.; Ruffieux, P.; Gröning, O.; Gröning, P.; Kastler, M.; Rieger, R.; Feng, X.; et al. Porous graphenes: Two-dimensional polymer synthesis with atomic precision. *Chem. Commun.* **2009**, 6919–6921. [CrossRef] [PubMed]

20. Blunt, M.O.; Russell, J.C.; Champness, N.R.; Beton, P.H. Templating molecular adsorption using a covalent organic framework. *Chem. Commun.* **2010**, *46*, 7157–7159. [CrossRef]

21. Sakamoto, J.; van Heijst, J.; Lukin, O.; Schluter, A.D. Two-dimensional polymers: Just a dream of synthetic chemists? *Angew. Chem. Int. Ed.* **2009**, *48*, 1030–1069. [CrossRef]

22. Geim, A.K. Graphene: Status and prospects. *Science* **2009**, *324*, 1530–1534. [CrossRef]

23. Mahmoudi, T.; Wang, Y.; Hahn, Y.-B. Graphene and its derivatives for solar cells application. *Nano Energy* **2018**, *47*, 51–65. [CrossRef]

24. Schwierz, F. Graphene transistors. *Nat. Nanotechnol.* **2010**, *5*, 487–496. [CrossRef]

25. Velasco-Soto, M.A.; Pérez-Garcıa, S.A.; Alvarez-Quintana, J.; Cao, Y.; Nyborg, L.; Licea-Jiménez, L. Selective band gap manipulation of graphene oxide by its reduction with mild reagents. *Carbon* **2015**, *93*, 967–973. [CrossRef]

26. Schedin, F.; Geim, A.K.; Morozov, S.V.; Hill, E.W.; Blake, P.; Katsnelson, M.I.; Novoselov, K.S. Detection of individual gas molecules adsorbed on graphene. *Nat. Mater.* **2007**, *6*, 652–655. [CrossRef]

27. Stepanow, S.; Lingenfelder, M.; Dmitriev, A.; Spillmann, H.; Delvigne, E.; Lin, N.; Deng, X.; Cai, C.; Barth, J.V.; Kern, K. Steering molecular organization and host-guest interactions using two-dimensional nanoporous coordination systems. *Nat. Mater.* **2004**, *3*, 229–233. [CrossRef] [PubMed]

28. Bieri, M.; Nguyen, M.T.; Gröning, O.; Cai, J.M.; Treier, M.; Ait-Mansour, K.; Ruffieux, P.; Pignedoli, C.A.; Passerone, D.; Kastler, M.; et al. Two-dimensional polymer formation on surfaces: Insight into the roles of precursor mobility and reactivity. *J. Am. Chem. Soc.* **2010**, *132*, 16669–16676. [CrossRef] [PubMed]

29. Ourdjini, O.; Pawlak, R.; Abel, M.; Clair, S.; Chen, L.; Bergeon, N.; Sassi, M.; Oison, V.; Debierre, J.-M.; Coratger, R.; et al. Substrate-mediated ordering and defect analysis of a surface covalent organic framework. *Phys. Rev. B* **2011**, *84*, 125421. [CrossRef]

30. Shen, Q.; Gao, H.-Y.; Fuchs, H. Frontiers of on-surface synthesis: From principles to applications. *Nano Today* **2017**, *13*, 77–96. [CrossRef]

31. Hu, J.; Liang, Z.; Shen, K.; Sun, H.; Jiang, Z.; Song, F. Recent progress in the fabrication of low dimensional nanostructures via surface-assisted transforming and coupling. *J. Nanomater.* **2017**, *2017*, 4796538. [CrossRef]

32. Treier, M.; Pignedoli, C.A.; Laino, T.; Rieger, R.; Müllen, K.; Passerone, D.; Fasel, R. Surface-assisted cyclodehydrogenation provides a synthetic route towards easily processable and chemically tailored nanographenes. *Nat. Chem.* **2011**, *3*, 61–67. [CrossRef]

33. Han, P.; Akagi, K.; Federici Canova, F.; Mutoh, H.; Shiraki, S.; Iwaya, K.; Weiss, P.S.; Asao, N.; Hitosugi, T. Bottom-up graphene-nanoribbon fabrication reveals chiral edges and enantioselectivity. *ACS Nano* **2014**, *8*, 9181–9187. [CrossRef]

34. Rastgoo-Lahrood, A.; Macknapp, K.; Ritter, V.; Sotier, S.; Heckl, W.M.; Lackinger, M.; Spitzer, S. Solvent-free on-surface synthesis of boroxine COF monolayers. *Chem. Commun.* **2017**, *53*, 5147–5150. [CrossRef]

35. Dienstmaier, J.F.; Medina, D.D.; Dogru, M.; Knochel, P.; Bein, T.; Heckl, W.M.; Lackinger, M. Isoreticular two-dimensional covalent organic frameworks synthesized by on-surface condensation of diboronic acids. *ACS Nano* **2012**, *6*, 7234–7242. [CrossRef]

36. Dienstmaier, J.F.; Gigler, A.M.; Goetz, A.J.; Knochel, P.; Bein, T.; Lyapin, A.; Reichlmaier, S.; Heckl, W.M.; Lackinger, M. Synthesis of well-ordered COF monolayers: Surface growth of nanocrystalline precursors versus direct on-surface polycondensation. *ACS Nano* **2011**, *5*, 9737–9745. [CrossRef]

37. Liu, C.; Zhang, W.; Zeng, Q.; Lei, S. A photoresponsive surface covalent organic framework: Surface-confined synthesis, isomerization, and controlled guest capture and release. *Chem. Eur. J.* **2016**, *22*, 6768–6773. [CrossRef]

38. Calik, M.; Sick, T.; Dogru, M.; Döblinger, M.; Datz, S.; Budde, H.; Hartschuh, A.; Auras, F.; Bein, T. From highly crystalline to outer surface-functionalized covalent organic frameworks: A modulation approach. *J. Am. Chem. Soc.* **2016**, *138*, 1234–1239. [CrossRef]

39. Zwaneveld, N.A.A.; Pawlak, R.; Abel, M.; Catalin, D.; Gigmes, D.; Bertin, D.; Porte, L. Organized formation of 2D extended covalent organic frameworks at surfaces. *J. Am. Chem. Soc.* **2008**, *130*, 6678–6679. [CrossRef]

40. Yu, L.; Li, Z.-B.; Wang, D. Construction of boronate ester based single-layered covalent organic frameworks. *Chem. Commun.* **2016**, *52*, 13771–13774. [CrossRef]

41. Sun, Q.; Zhang, C.; Li, Z.; Kong, H.; Tan, Q.; Hu, A.; Xu, W. On-surface formation of one-dimensional polyphenylene through Bergman cyclization. *J. Am. Chem. Soc.* **2013**, *135*, 8448–8451. [CrossRef] [PubMed]

42. Gao, H.-Y.; Wagner, H.; Zhong, D.; Franke, J.-H.; Studer, A.; Fuchs, H. Glaser coupling at metal surfaces. *Angew. Chem. Int. Ed.* **2013**, *52*, 4024–4028. [CrossRef] [PubMed]

43. Klappenberger, F.; Hellwig, R.; Du, P.; Paintner, T.; Uphoff, M.; Zhang, L.; Lin, T.; Moghanaki, B.A.; Paszkiewicz, M.; Vobornik, I.; et al. Functionalized graphdiyne nanowires: On-surface synthesis and assessment of band structure, flexibility, and information storage potential. *Small* **2018**, *14*, 1704321. [CrossRef] [PubMed]

44. Sun, Q.; Cai, L.; Ding, Y.; Ma, H.; Yuan, C.; Xu, W. Single-molecule insight into Wurtz reactions on metal surfaces. *Phys. Chem. Chem. Phys.* **2016**, *18*, 2730–2735. [CrossRef] [PubMed]

45. Sun, Q.; Cai, L.; Ding, Y.; Xie, L.; Zhang, C.; Tan, Q.; Xu, W. Dehydrogenative homocoupling of terminal alkenes on copper surfaces: A route to dienes. *Angew. Chem. Int. Ed.* **2015**, *54*, 4549–4552. [CrossRef] [PubMed]

46. Wiengarten, A.; Seufert, K.; Auwärter, W.; Ecija, D.; Diller, K.; Allegretti, F.; Bischoff, F.; Fischer, S.; Duncan, D.A.; Papageorgiou, A.C.; et al. Surface-assisted dehydrogenative homocoupling of porphine molecules. *J. Am. Chem. Soc.* **2014**, *136*, 9346–9354. [CrossRef] [PubMed]

47. Sun, Q.; Zhang, C.; Kong, H.; Tan, Q.; Xu, W. On-surface aryl-aryl coupling via selective C-H activation. *Chem. Commun.* **2014**, *50*, 11825–11828. [CrossRef]

48. Shu, C.-H.; Liu, M.-X.; Zha, Z.-Q.; Pan, J.-L.; Zhang, S.-Z.; Xie, Y.-L.; Chen, J.-L.; Yuan, D.-W.; Qiu, X.-H.; Liu, P.-N. On-surface synthesis of poly(p-phenylene ethynylene) molecular wires via in situ formation of carbon-carbon triple bond. *Nat. Commun.* **2018**, *9*, 2322. [CrossRef]

49. Yang, B.; Björk, J.; Lin, H.; Zhang, X.; Zhang, H.; Li, Y.; Fan, J.; Li, Q.; Chi, L. Synthesis of surface covalent organic frameworks via dimerization and cyclotrimerization of acetyls. *J. Am. Chem. Soc.* **2015**, *137*, 4904–4907. [CrossRef]

50. Weigelt, S.; Busse, C.; Bombis, C.; Knudsen, M.M.; Gothelf, K.V.; Strunskus, T.; Wöll, C.; Dahlbom, M.; Hammer, B.; Lægsgaard, E.; et al. Covalent interlinking of an aldehyde and an amine on a Au(111) surface in ultrahigh vacuum. *Angew. Chem. Int. Ed.* **2007**, *46*, 9227–9230. [CrossRef]

51. Weigelt, S.; Busse, C.; Bombis, C.; Knudsen, M.M.; Gothelf, K.V.; Lægsgaard, E.; Besenbacher, F.; Linderoth, T.R. Surface synthesis of 2D branched polymer nanostructures. *Angew. Chem. Int. Ed.* **2008**, *47*, 4406–4410. [CrossRef] [PubMed]

52. Tanoue, R.; Higuchi, R.; Enoki, N.; Miyasato, Y.; Uemura, S.; Kimizuka, N.; Stieg, A.Z.; Gimzewski, J.K.; Kunitake, M. Thermodynamically controlled self-assembly of covalent nanoarchitectures in aqueous solution. *ACS Nano* **2011**, *5*, 3923–3929. [CrossRef] [PubMed]

53. Li, Y.B.; Wan, J.H.; Deng, K.; Han, X.N.; Lei, S.B.; Yang, Y.L.; Zheng, Q.Y.; Zeng, Q.D.; Wang, C. Transformation of self-assembled structure by the addition of active reactant. *J. Phys. Chem. C* **2011**, *115*, 6540–6544. [CrossRef]

54. Tanoue, R.; Higuchi, R.; Ikebe, K.; Uemura, S.; Kimizuka, N.; Stieg, A.Z.; Gimzewski, J.K.; Kunitake, M. In situ STM investigation of aromatic poly(azomethine) arrays constructed by "on-site" equilibrium polymerization. *Langmuir* **2012**, *28*, 13844–13851. [CrossRef] [PubMed]

55. Liu, X.H.; Guan, C.Z.; Ding, S.Y.; Wang, W.; Yan, H.J.; Wang, D.; Wan, L.J. On-surface synthesis of single-layered two-dimensional covalent organic frameworks via solid-vapor interface reactions. *J. Am. Chem. Soc.* **2013**, *135*, 10470–10474. [CrossRef] [PubMed]

56. Hu, F.Y.; Zhang, X.M.; Wang, X.; Wang, S.; Wang, H.Q.; Duan, W.B.; Zeng, Q.D.; Wang, C. In situ STM investigation of two-dimensional chiral assemblies through Schiff-base condensation at a liquid/solid interface. *ACS Appl. Mater. Interf.* **2013**, *5*, 1583–1587. [CrossRef] [PubMed]

57. Xu, L.R.; Zhou, X.; Yu, Y.X.; Tian, W.Q.; Ma, J.; Lei, S.B. Surface-confined crystalline two-dimensional covalent organic frameworks. *ACS Nano* **2013**, *7*, 8066–8073. [CrossRef] [PubMed]

58. Bisbey, R.P.; Dichtel, W.R. Covalent organic frameworks as a platform for multidimensional polymerization. *ACS Cent. Sci.* **2017**, *3*, 533–543. [CrossRef]

59. Marele, A.C.; Mas-Ballesté, R.; Terracciano, L.; Rodríguez-Fernández, J.; Berlanga, I.; Alexandre, S.S.; Otero, R.; Gallego, J.M.; Zamora, F.; Gómez-Rodríguez, J.M. Formation of a surface covalent organic framework based on polyester condensation. *Chem. Commun.* **2012**, *48*, 6779–6781. [CrossRef] [PubMed]

60. Treier, M.; Richardson, N.V.; Fasel, R. Fabrication of surface-supported low-dimensional polyimide networks. *J. Am. Chem. Soc.* **2008**, *130*, 14054–14055. [CrossRef] [PubMed]

61. Treier, M.; Fasel, R.; Champness, N.R.; Argent, S.; Richardson, N.V. Molecular imaging of polyimide formation. *Phys. Chem. Chem. Phys.* **2009**, *11*, 1209–1214. [CrossRef] [PubMed]

62. Jensen, S.; Greenwood, J.; Früchtl, H.A.; Baddeley, C.J. STM investigation on the formation of oligoamides on Au{111} by surface-confined reactions of melamine with trimesoyl chloride. *J. Phys. Chem. C* **2011**, *115*, 8630–8636. [CrossRef]

63. Schmitz, C.H.; Ikonomov, J.; Sokolowski, M. Two-dimensional polyamide networks with a broad pore size distribution on the Ag(111) surface. *J. Phys. Chem. C* **2011**, *115*, 7270–7278. [CrossRef]

64. Wäckerlin, C.; Li, J.; Mairena, A.; Martin, K.; Avarvari, N.; Ernst, K.-H. Surface-assisted diastereoselective Ullmann coupling of bishelicenes. *Chem. Commun.* **2016**, *52*, 12694–12697. [CrossRef] [PubMed]

65. Kalashnyk, N.; Mouhat, K.; Oh, J.; Jung, J.; Xie, Y.; Giovanelli, L.; Salomon, E.; Angot, T.; Dumur, F.; Gigmes, D.; Clair, S. On-surface synthesis of aligned functional nanoribbons monitored by vibrational spectroscopy. *Nature Commun.* **2017**, *8*, 14735. [CrossRef] [PubMed]

66. Kalashnyk, N.; Salomon, E.; Mun, S.H.; Jung, J.; Giovanelli, L.; Angot, T.; Dumur, F.; Gigmes, D.; Clair, S. The orientation of silver surfaces drives the reactivity and the selectivity in homo-coupling reactions. *ChemPhysChem* **2018**, *19*, 1802–1808. [CrossRef] [PubMed]

67. Kalashnyk, N.; Dumur, F.; Gigmes, D.; Clair, S. Molecular adaptation in supra-molecular self-assembly: Brickwall-type phases of indacene-tetrone on silver surfaces. *Chem. Commun.* **2018**, *54*, 8510–8513. [CrossRef]

68. Zhong, D.; Franke, J.-H.; Podiyanachari, S.K.; Blömker, T.; Zhang, H.; Kehr, G.; Erker, G.; Fuchs, H.; Chi, L. Linear alkane polymerization on a gold surface. *Science* **2011**, *334*, 213–216. [CrossRef]

69. In't Veld, M.; Iavicoli, P.; Haq, S.; Amabilino, D.B.; Raval, R. Unique intermolecular reaction of simple porphyrins at a metal surface gives covalent nanostructures. *Chem. Commun.* **2008**, 1536–1538. [CrossRef]

70. Held, P.A.; Fuchs, H.; Studer, A. Covalent-bond formation via on-surface chemistry. *Chem. Eur. J.* **2017**, *23*, 5874–5892. [CrossRef]

71. Dienel, T.; Gomez-Diaz, J.; Seitsonen, A.P.; Widmer, R.; Iannuzzi, M.; Radican, K.; Sachdev, H.; Mullen, K.; Hutter, J.; Groning, O. Dehalogenation and coupling of a polycyclic hydrocarbon on an atomically thin insulator. *ACS Nano* **2014**, *8*, 6571–6579. [CrossRef]

72. Killops, K.L.; Campos, L.M.; Hawker, C.J. Robust, efficient, and orthogonal synthesis of dendrimers via thiol-ene "Click" chemistry. *J. Am. Chem. Soc.* **2008**, *130*, 5062–5064. [CrossRef]

73. Fickert, J.; Makowski, M.; Kappl, M.; Landfester, K.; Crespy, D. Efficient encapsulation of self-healing agents in polymer nanocontainers functionalized by orthogonal reactions. *Macromolecules* **2012**, *45*, 6324–6332. [CrossRef]

74. Joralemon, M.J.; O'Reilly, R.K.; Hawker, C.J.; Wooley, K.L. Shell click-crosslinked (SCC) nanoparticles: A new methodology for synthesis and orthogonal functionalization. *J. Am. Chem. Soc.* **2005**, *127*, 16892–16899. [CrossRef]

75. Zeng, Y.-F.; Zou, R.-Y.; Luo, Z.; Zhang, H.-C.; Yao, X.; Ma, X.; Zou, R.-Q.; Zhao, Y.-L. Covalent organic frameworks formed with two types of covalent bonds based on orthogonal reactions. *J. Am. Chem. Soc.* **2015**, *137*, 1020–1023. [CrossRef] [PubMed]

76. Chen, X.; Addicoat, M.; Jin, E.-Q.; Xu, H.; Hayashi, T.; Xu, F.; Huang, N.; Irle, S.; Jiang, D.-L. Designed synthesis of double-stage two-dimensional covalent organic frameworks. *Sci. Rep.* **2015**, *5*, 14650–14668. [CrossRef] [PubMed]

77. Moreno, C.; Vilas-Varela, M.; Kretz, B.; Garcia-Lekue, A.; Costache, M.V.; Paradinas, M.; Panighel, M.; Ceballos, G.; Valenzuela, S.O.; Peña, D.; et al. Bottom-up synthesis of multifunctional nanoporous graphene. *Science* **2018**, *360*, 199–203. [CrossRef] [PubMed]

78. Chen, C.; Joshi, T.; Li, H.; Chavez, A.D.; Pedramrazi, Z.; Liu, P.-N.; Li, H.; Dichtel, W.R.; Bredas, J.-L.; Crommie, M.F. Local electronic structure of a single-layer porphyrin-containing covalent organic framework. *ACS Nano* **2018**, *12*, 385–391. [CrossRef] [PubMed]

79. Ruiz-Oses, M.; de Oteyza, D.G.; Fernandez-Torrente, I.; Gonzalez-Lakunza, N.; Schmidt-Weber, P.M.; Kampen, T.; Horn, K.; Gourdon, A.; Arnau, A.; Ortega, J.E. Non-covalent interactions in supramolecular assemblies investigated with electron spectroscopies. *Chemphyschem* **2009**, *10*, 896–900. [CrossRef] [PubMed]

80. Gambardella, P.; Stepanow, S.; Dmitriev, A.; Honolka, J.; de Groot, F.M.F.; Lingenfelder, M.; Sen Gupta, S.; Sarma, D.D.; Bencok, P.; Stanescu, S.; et al. Supramolecular control of the magnetic anisotropy in two-dimensional high-spin Fe arrays at a metal interface. *Nat. Mater.* **2009**, *8*, 189–193. [CrossRef] [PubMed]

81. Fakirov, S. Condensation polymers: Their chemical peculiarities offer great opportunities. *Progr. Polym. Sci.* **2019**, *89*, 1–18. [CrossRef]

82. Lipton-Duffin, J.A.; Ivasenko, O.; Perepichka, D.F.; Rosei, F. Synthesis of polyphenylene molecular wires by surface-confined polymerization. *Small* **2009**, *5*, 592–597. [CrossRef]

83. Lafferentz, L.; Eberhardt, V.; Dri, C.; Africh, C.; Comelli, G.; Esch, F.; Hecht, S.; Grill, L. Controlling on-surface polymerization by hierarchical and substrate-directed growth. *Nat. Chem.* **2012**, *4*, 215–220. [CrossRef] [PubMed]

84. Schlögl, S.; Heckl, W.M.; Lackinger, M. On-surface radical addition of triply iodinated monomers on Au(111): The influence of monomer size and thermal post-processing. *Surf. Sci.* **2012**, *606*, 999–1004. [CrossRef]

85. Walch, H.; Gutzler, R.; Sirtl, T.; Eder, G.; Lackinger, M. Material- and orientation-dependent reactivity for heterogeneously catalyzed carbon-bromine bond homolysis. *J. Phys. Chem. C* **2010**, *114*, 12604–12609. [CrossRef]

86. Gutzler, R.; Walch, H.; Eder, G.; Kloft, S.; Heckl, W.M.; Lackinger, M. Surface mediated synthesis of 2D covalent organic frameworks: 1,3,5-tris(4-bromophenyl)benzene on Graphite(001), Cu(111), and Ag(110). *Chem. Commun.* **2009**, 4456–4458. [CrossRef] [PubMed]

87. Gutzler, R.; Cardenas, L.; Lipton-Duffin, J.; El Garah, M.; Dinca, L.E.; Szakacs, C.E.; Fu, C.; Gallagher, M.; Vondracek, M.; Rybachuk, M.; et al. Ullmann-type coupling of brominated tetrathienoanthracene on copper and silver. *Nanoscale* **2014**, *6*, 2660–2668. [CrossRef] [PubMed]

88. Eichhorn, J.; Strunskus, T.; Rastgoo-Lahrood, A.; Samanta, D.; Schmittel, M.; Lackinger, M. On-surface Ullmann polymerization via intermediate organometallic networks on Ag(111). *Chem. Commun.* **2014**, *50*, 7680–7682. [CrossRef]

89. Eichhorn, J.; Nieckarz, D.; Ochs, O.; Samanta, D.; Schmittel, M.; Szabelski, P.J.; Lackinger, M. On-surface Ullmann coupling: The influence of kinetic reaction parameters on the morphology and quality of covalent networks. *ACS Nano* **2014**, *8*, 7880–7889. [CrossRef]

90. Shi, K.J.; Zhang, X.; Shu, C.H.; Li, D.Y.; Wu, X.Y.; Liu, P.N. Ullmann coupling reaction of aryl chlorides on Au(111) using dosed Cu as a catalyst and the programmed growth of 2D covalent organic frameworks. *Chem. Commun.* **2016**, *52*, 8726–8729. [CrossRef]

91. Peyrot, D.; Silly, M.G.; Silly, F. Temperature-triggered sequential on-surface synthesis of one and two covalently bonded porous organic nanoarchitectures on Au(111). *J. Phys. Chem. C* **2017**, *121*, 26815–26821. [CrossRef]

92. Guan, C.Z.; Wang, D.; Wan, L.J. Construction and repair of highly ordered 2D covalent networks by chemical equilibrium regulation. *Chem. Commun.* **2012**, *48*, 2943–2945. [CrossRef] [PubMed]

93. Stredansky, M.; Sala, A.; Fontanot, T.; Costantini, R.; Africh, C.; Comelli, G.; Floreano, L.; Morgante, A.; Cossaro, A. On-surface synthesis of a 2D boroxine framework: A route to a novel 2D material? *Chem. Commun.* **2018**, *54*, 3971–3973. [CrossRef] [PubMed]

94. Schlögl, S.; Sirtl, T.; Eichhorn, J.; Heckl, W.M.; Lackinger, M. Synthesis of two-dimensional phenylene–boroxine networks through in vacuo condensation and on-surface radical addition. *Chem. Commun.* **2011**, *47*, 12355–12357. [CrossRef]

95. Faury, T.; Clair, S.; Abel, M.; Dumur, F.; Gigmes, D.; Porte, L. Sequential linking to control growth of a surface covalent organic framework. *J. Phys. Chem. C* **2012**, *116*, 4819–4823. [CrossRef]

96. Clair, S.; Ourdjini, O.; Abel, M.; Porte, L. Tip- or electron beam-induced surface polymerization. *Chem. Commun.* **2011**, *47*, 8028–8030. [CrossRef]

97. Yue, J.-Y.; Mo, Y.-P.; Li, S.-Y.; Dong, W.-L.; Chen, T.; Wang, D. Simultaneous construction of two linkages for the on-surface synthesis of imine–boroxine hybrid covalent organic frameworks. *Chem. Sci.* **2017**, *8*, 2169–2174. [CrossRef]

98. Koudia, M.; Abel, M. Step-by-step on-surface synthesis: From manganese phthalocyanines to their polymeric form. *Chem. Commun.* **2014**, *50*, 8565–8567. [CrossRef]

99. Gottfried, J.M. Surface chemistry of porphyrins and phthalocyanines. *Surf. Sci. Rep.* **2015**, *70*, 259–379. [CrossRef]

100. Abel, M.; Clair, S.; Ourdjini, O.; Mossoyan, M.; Porte, L. Single layer of polymeric Fe-phthalocyanine: An organometallic sheet on metal and thin insulating film. *J. Am. Chem. Soc.* **2011**, *133*, 1203–1205. [CrossRef]

101. Cai, J.; Ruffieux, P.; Jaafar, R.; Bieri, M.; Braun, T.; Blankenburg, S.; Muoth, M.; Seitsonen, A.P.; Saleh, M.; Feng, X.; et al. Atomically precise bottom-up fabrication of graphene nanoribbons. *Nature* **2010**, *466*, 470–474. [CrossRef] [PubMed]

102. Teeter, J.D.; Costa, P.S.; Zahl, P.; Vo, T.H.; Shekhirev, M.; Xu, W.; Zeng, X.C.; Enders, A.; Sinitskii, A. Dense monolayer films of atomically precise graphene nanoribbons on metallic substrates enabled by direct contact transfer of molecular precursors. *Nanoscale* **2017**, *9*, 18835–18844. [CrossRef] [PubMed]

103. Kimouche, A.; Ervasti, M.M.; Drost, R.; Halonen, S.; Harju, A.; Joensuu, P.M.; Sainio, J.; Liljeroth, P. Ultra-narrow metallic armchair graphene nanoribbons. *Nat. Commun.* **2015**, *6*, 10177. [CrossRef] [PubMed]

104. Abdurakhmanova, N.; Amsharov, N.; Stepanow, S.; Jansen, M.; Kern, K.; Amsharov, K. Synthesis of wide atomically precise graphene nanoribbons from para-oligophenylene based molecular precursor. *Carbon* **2014**, *77*, 1187–1190. [CrossRef]

105. Han, P.; Akagi, K.; Federici Canova, F.; Shimizu, R.; Oguchi, H.; Shiraki, S.; Weiss, P.S.; Asao, N.; Hitosugi, T. Self-assembly strategy for fabricating connected graphene nanoribbons. *ACS Nano* **2015**, *9*, 12035–12044. [CrossRef] [PubMed]

106. Ruffieux, P.; Wang, S.; Yang, B.; Sánchez-Sánchez, C.; Liu, J.; Dienel, T.; Talirz, L.; Shinde, P.; Pignedoli, C.A.; Passerone, D.; et al. On-surface synthesis of graphene nanoribbons with zigzag edge topology. *Nature* **2016**, *531*, 489–492. [CrossRef] [PubMed]

107. Simonov, K.A.; Vinogradov, N.A.; Vinogradov, A.S.; Generalov, A.V.; Zagrebina, E.M.; Svirskiy, G.I.; Cafolla, A.A.; Carpy, T.; Cunniffe, J.P.; Taketsugu, T.; et al. From graphene nanoribbons on Cu(111) to nanographene on Cu(110): Critical role of substrate structure in the bottom-up fabrication strategy. *ACS Nano* **2015**, *9*, 8997–9011. [CrossRef] [PubMed]

108. Sánchez-Sánchez, C.; Dienel, T.; Deniz, O.; Ruffieux, P.; Berger, R.; Feng, X.; Müllen, K.; Fasel, R. Purely armchair or partially chiral: Noncontact atomic force microscopy characterization of dibromobianthryl-based graphene nanoribbons grown on Cu(111). *ACS Nano* **2016**, *10*, 8006–8011. [CrossRef] [PubMed]

109. Talirz, L.; Söde, H.; Dumslaff, T.; Wang, S.; Sanchez-Valencia, J.R.; Liu, J.; Shinde, P.; Pignedoli, C.A.; Liang, L.; Meunier, V.; et al. On-surface synthesis and characterization of 9-atom wide armchair graphene nanoribbons. *ACS Nano* **2017**, *11*, 1380–1388. [CrossRef] [PubMed]

110. Teeter, J.D.; Costa, P.S.; Mehdi Pour, M.; Miller, D.P.; Zurek, E.; Enders, A.; Sinitskii, A. Epitaxial growth of aligned atomically precise chevron graphene nanoribbons on Cu(111). *Chem. Commun.* **2017**, *53*, 8463–8466. [CrossRef] [PubMed]

111. Bronner, C.; Stremlau, S.; Gille, M.; Brauße, F.; Haase, A.; Hecht, S.; Tegeder, P. Aligning the band gap of graphene nanoribbons by monomer doping. *Angew. Chem. Int. Ed.* **2013**, *52*, 4422–4425. [CrossRef] [PubMed]

112. Vo, T.H.; Perera, U.G.E.; Shekhirev, M.; Mehdi Pour, M.; Kunkel, D.A.; Lu, H.; Gruverman, A.; Sutter, E.; Cotlet, M.; Nykypanchuk, D.; et al. Nitrogen-doping induced self-assembly of graphene nanoribbon-based two-dimensional and three-dimensional metamaterials. *Nano Lett.* **2015**, *15*, 5770–5777. [CrossRef]

113. Cloke, R.R.; Marangoni, T.; Nguyen, G.D.; Joshi, T.; Rizzo, D.J.; Bronner, C.; Cao, T.; Louie, S.G.; Crommie, M.F.; Fischer, F.R. Site-specific substitutional boron doping of semiconducting armchair graphene nanoribbons. *J. Am. Chem. Soc.* **2015**, *137*, 8872–8875. [CrossRef] [PubMed]

114. Kawai, S.; Saito, S.; Osumi, S.; Yamaguchi, S.; Foster, A.S.; Spijker, P.; Meyer, E. Atomically controlled substitutional boron-doping of graphene nanoribbons. *Nat. Commun.* **2015**, *6*, 8098. [CrossRef] [PubMed]

115. Marangoni, T.; Haberer, D.; Rizzo, D.J.; Cloke, R.R.; Fischer, F.R. Heterostructures through divergent edge reconstruction in nitrogen-doped segmented graphene nanoribbons. *Chem. Eur. J.* **2016**, *22*, 13037–13040. [CrossRef]

116. Cai, J.; Pignedoli, C.A.; Talirz, L.; Ruffieux, P.; Soede, H.; Liang, L.; Meunier, V.; Berger, R.; Li, R.; Feng, X.; et al. Graphene nanoribbon heterojunctions. *Nat. Nanotechnol.* **2014**, *9*, 896–900. [CrossRef] [PubMed]

117. Chen, Y.-C.; Cao, T.; Chen, C.; Pedramrazi, Z.; Haberer, D.; de Oteyza, D.G.; Fischer, F.R.; Louie, S.G.; Crommie, M.F. Molecular bandgap engineering of bottom up synthesized graphene nanoribbon heterojunctions. *Nat. Nanotechnol.* **2015**, *10*, 156–160. [CrossRef] [PubMed]

118. Zhang, H.; Lin, H.; Sun, K.; Chen, L.; Zagranyarski, Y.; Aghdassi, N.; Duhm, S.; Li, Q.; Zhong, D.; Li, Y.; et al. On-surface synthesis of rylene-type graphene nanoribbons. *J. Am. Chem. Soc.* **2015**, *137*, 4022–4025. [CrossRef] [PubMed]

119. de Oteyza, D.G.; García-Lekue, A.; Vilas-Varela, M.; Merino-Díez, N.; Carbonell-Sanromà, E.; Corso, M.; Vasseur, G.; Rogero, C.; Guitián, E.; Pascual, J.I.; et al. Substrate-independent growth of atomically precise chiral graphene nanoribbons. *ACS Nano* **2016**, *10*, 9000–9008. [CrossRef] [PubMed]

120. Creencia, E.C.; Horaguchi, T. Thermal decomposition reactions of n-alkylated 2-aminobiphenyls to carbazole and phenanthridine. *J. Heterocycl. Chem.* **2006**, *43*, 1441–1446. [CrossRef]

121. Schulz, F.; Jacobse, P.H.; Canova, F.F.; van der Lit, J.; Gao, D.Z.; van den Hoogenband, A.; Han, P.; Klein Gebbink, R.J.M.; Moret, M.-E.; Joensuu, P.M.; et al. Precursor geometry determines the growth mechanism in graphene nanoribbons. *J. Phys. Chem. C* **2017**, *121*, 2896–2904. [CrossRef]

122. Simonov, K.A.; Vinogradov, N.A.; Vinogradov, A.S.; Generalov, A.V.; Zagrebina, E.M.; Mårtensson, N.; Cafolla, A.A.; Carpy, T.; Cunniffe, J.P.; Preobrajenski, A.B. Comment on "Bottom-up graphene-nanoribbon fabrication reveals chiral edges and enantioselectivity". *ACS Nano* **2015**, *9*, 3399–3403. [CrossRef] [PubMed]

123. Han, P.; Akagi, K.; Federici Canova, F.; Mutoh, H.; Shiraki, S.; Iwaya, K.; Weiss, P.S.; Asao, N.; Hitosugi, T. Reply to "Comment on 'Bottom-up graphene-nanoribbon fabrication reveals chiral edges and enantioselectivity'". *ACS Nano* **2015**, *9*, 3404–3405. [CrossRef] [PubMed]

124. Liu, J.; Chen, Q.; He, Q.; Zhang, Y.; Fu, X.; Wang, X.; Zhao, D.; Chen, W.; Xu, G.Q.; Wu, K. Bromine adatom promoted C–H bond activation in terminal alkynes at room temperature on Ag(111). *Phys. Chem. Chem. Phys.* **2018**, *20*, 11081–11088. [CrossRef] [PubMed]

125. Held, P.A.; Gao, H.-Y.; Liu, L.; Mîck-Lichtenfeld, C.; Timmer, A.; Mönig, H.; Barton, D.; Neugebauer, J.; Fuchs, H.; Studer, A. On-surface domino reactions: Glaser coupling and dehydrogenative coupling of a biscarboxylic acid to form polymeric bisacylperoxides. *Angew. Chem. Int. Ed.* **2016**, *55*, 9777–9782. [CrossRef] [PubMed]

126. Clair, S.; Pons, S.; Seitsonen, A.P.; Brune, H.; Kern, K.; Barth, J.V. STM study of terephthalic acid self-assembly on Au(111): Hydrogen-bonded sheets on an inhomogeneous substrate. *J. Phys. Chem. B* **2004**, *108*, 14585–14590. [CrossRef]

127. Payer, D.; Comisso, A.; Dmitriev, A.; Strunskus, T.; Lin, N.; Wöll, C.; DeVita, A.; Barth, J.V.; Kern, K. Ionic hydrogen bonds controlling two-dimensional supramolecular systems at a metal surface. *Chem. Eur. J.* **2007**, *13*, 3900–3906. [CrossRef] [PubMed]

128. Dmitriev, A.; Lin, N.; Weckesser, J.; Barth, J.V.; Kern, K. Supramolecular assemblies of trimesic acid on a Cu(100) Surface. *J. Phys. Chem. B* **2002**, *106*, 6907–6912. [CrossRef]

129. Gao, H.-Y.; Held, P.A.; Knor, M.; Mîck-Lichtenfeld, C.; Neugebauer, J.; Studer, A.; Fuchs, H. Decarboxylative polymerization of 2,6-naphthalenedicarboxylic acid at surfaces. *J. Am. Chem. Soc.* **2014**, *136*, 9658–9663. [CrossRef] [PubMed]

Article

Old Molecule, New Chemistry: Exploring Silicon Phthalocyanines as Emerging N-Type Materials in Organic Electronics

Nathan J. Yutronkie [1], Trevor M. Grant [2], Owen A. Melville [2], Benoît H. Lessard [2,*] and Jaclyn L. Brusso [1,*]

[1] Department of Chemistry and Biomolecular Sciences, University of Ottawa, 150 Louis Pasteur, Ottawa, ON K1N 6N5, Canada; nyutr055@uottawa.ca

[2] Department of Chemical and Biological Engineering, University of Ottawa, 161 Louis Pasteur, Ottawa, ON K1N 6N5, Canada; tgran079@uottawa.ca (T.M.G.); omelv065@uottawa.ca (O.A.M.)

* Correspondence: benoit.lessard@uottawa.ca (B.H.L.); jbrusso@uottawa.ca (J.L.B.)

Received: 10 April 2019; Accepted: 20 April 2019; Published: 24 April 2019

Abstract: Efficient synthesis of silicon phthalocyanines (SiPc) eliminating the strenuous reaction conditions and hazardous reagents required by classical methods is described. Implementation into organic thin-film transistors (OTFTs) affords average electron field-effect mobility of 3.1×10^{-3} cm^2 V^{-1} s^{-1} and threshold voltage of 25.6 V for all synthetic routes. These results demonstrate that our novel chemistry can lead to high performing SiPc-based n-type OTFTs.

Keywords: silicon phthalocyanines; n-type organic semiconductors; organic thin-film transistors

1. Introduction

In recent years, research focused on the design of various organic architectures that enable electron transport in organic semiconductors (OSCs) has led to significant advances, not only in terms of OSC design but also the requirements for integration into organic electronics. In regard to the latter, to be viable for organic thin-film transistors (OTFTs), OSCs must ultimately possess high field-effect mobilities, robust environmental stability and ease of processability. While these efforts have led to impressive progress, the development of n-type OSCs still lags behind their p-type counterparts in terms of carrier mobility and air stability [1,2]. This may be attributed to the challenges associated with designing new n-type OSCs, in which the number of electron-deficient π-building blocks available is limited, and even fewer have been demonstrated to lead to high-mobilities [1]. Nonetheless, by employing motifs such as imides, carbonyls, *N*-heterocycles, cyanoethylene, and fullerenes, a library of both molecular and polymeric n-type OSCs has been achieved.

Phthalocyanines (Pcs; Figure 1), which are conjugated molecules composed of a tetramer of nitrogen-linked isoindole units, are ideal candidates for organic electronics. This is due in large part to their unique photophysical signatures and electrophysical properties, which are readily tunable through structural modification. Furthermore, they are known for their chemical robustness, high thermal resistance and low photodegradation [3,4]. Pcs often chelate a metal or metalloid through four M–N bonds, resulting in highly stable materials that have been explored for a variety of applications including their employment as active materials in organic electronics, predominantly as p-type materials [3,5–10]. With respect to their potential as n-type OSCs, only a handful of metal phthalocyanines (MPcs) have been shown to possess air stable n-type mobilities when implemented into OTFTs, with values on the order of 10^{-1} cm^{2}V^{-1}s^{-1} [9,11–13].

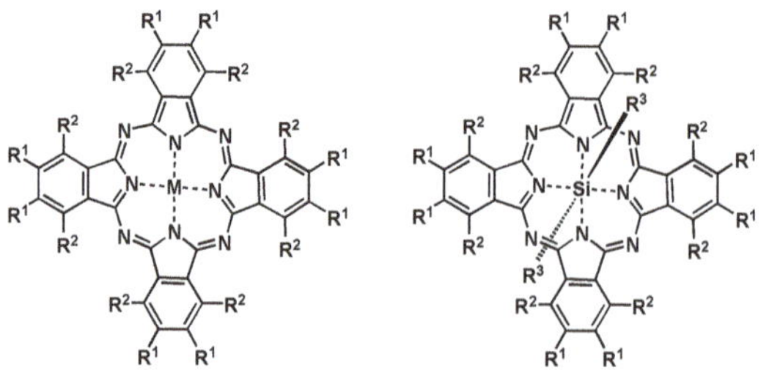

Figure 1. Divalent metal phthalocyanine (MPc) and tetravalent silicon phthalocyanine (SiPc) frameworks.

Replacing the central metal ion in MPcs with silicon affords tetravalent silicon phthalocyanines (SiPc), a subclass of phthalocyanines that have recently experienced an increase in interest with respect to integration into organic light emitting diodes [14–16], organic solar cells [17–21] and OTFTs [22], positioning SiPcs as excellent candidates as n-type OSCs. In addition to their inherent n-type mobility, a key advantage to employing SiPcs over divalent MPcs is the ability to modify the axial substituents (e.g., R^3), thereby providing a facile avenue for functionalization. Considering that the two key factors that determine the performance of OSCs in organic electronics are the frontier molecular orbitals (in terms of both energy levels and electronic distribution), and molecular packing in the solid state, axial substitution in SiPcs enables modification, and ideally control, of the film morphology. To that end, preliminary studies on SiPcs axially functionalized with aromatic substituents revealed electron mobilities up to 10^{-2} cm^{2}V^{-1}s^{-1}, demonstrating the influence of film morphology on device performance [22]. While axial substitution impacts molecular packing in the solid state, it has little effect on the energy levels of the frontier molecular orbitals. To that end, modifications to the R^1 and R^2 substituents about the Pc framework will not only influence the packing but also the energetics. This would therefore require appropriately substituted starting materials.

While MPcs are often prepared through the reaction of phthalonitrile with various metal precursors (e.g., MX$_2$, where X = halide), this route is not effective for silicon-based derivatives as very low yields (i.e., less than 1%) [23] of SiPcCl$_2$ are achieved upon the treatment of phthalonitrile with silicon tetrachloride (SiCl$_4$) [24]. Other reported attempts to prepare SiPcCl$_2$ involved the reaction of the disodium salt of phthalocyanine with SiCl$_4$; however, no product was obtained through this methodology [25]. Alternatively, starting with diiminoisoindoline or o-cyanobenzamide in place of phthalonitrile affords enhanced yields of 71% and 35%, respectively [26]. Although this improved the yield of SiPcCl$_2$, the preparation of such reagents proves problematic, especially of diiminoisoindoline (DIII). For example, the reported synthesis involves a constant flow of ammonia gas bubbling though a refluxing solution of phthalonitrile and sodium methoxide [26]. In addition to the harsh reaction conditions, the isolation of DIII of high purity is difficult to achieve, which may be attributed to many factors including its high solubility in a number of solvents, its tendency to cyclize upon dissolution and its sensitivity towards heat. This therefore limits the methods available for purification. As a result, DIII is often used in its crude form in most cases, or used directly in a one-pot synthesis of SiPcCl$_2$ [27]. In order to tap into the potential of SiPcs, we sought to explore and develop new synthetic protocols to facilitate advancements within this exciting class of n-type OSCs. To that end, we herein present effective and alternative approaches to the classical preparative route for SiPcCl$_2$, eliminating the undesired expenses attributed to the strenuous reaction conditions required and the hazardous reagents employed. Furthermore, integration of the resulting SiPc derivatives into high performing n-type OTFTs is presented, exhibiting the highest electron field-effect mobilities (μ_e) reported for silicon phthalocyanine-based OTFTs.

2. Results and Discussion

As mentioned, SiPcs can be isolated following Route A (Scheme 1); however, the conditions required are not appealing for large scale production due to the safety precautions necessary when dealing with corrosive gases and highly reducing sodium metal. To achieve a more versatile and less hazardous preparative method, we sought to eliminate the necessity of both sodium metal and ammonia gas in the preparation of DIII. To that end, utilization of lithium bis(trimethylsilyl)amide etherate (LiN(TMS)$_2$·Et$_2$O) offers several advantages over the reagents required for Route A. First and foremost, LiN(TMS)$_2$·Et$_2$O is a solid tolerant to atmospheric conditions, rendering it easier to handle and quantify than gaseous reagents such as ammonia. Not only does the use of LiN(TMS)$_2$·Et$_2$O eliminate the need for ammonia gas, but it also affords the necessary basicity without the use of sodium metal, thereby affording a preparative route that removes the use of toxic gases (ammonia) and highly reactive reagents (sodium metal). As outlined in Scheme 1, treatment of phthalonitrile with LiN(TMS)$_2$·Et$_2$O affords the anionic intermediate 1. Initial attempts to quench with either water or methanol led to various intractable by-products, likely due to decomposition of the desired product. Alternatively, quenching the anionic intermediate with gaseous hydrochloric acid afforded the dihydrogen chloride salt [H$_2$DIII]Cl$_2$ (Route B), which precipitates out of solution as a powdery yellow solid. Attempts to deprotonate [H$_2$DIII]Cl$_2$ to afford DIII were unsuccessful; however, preparation of SiPcCl$_2$ could be achieved by reacting [H$_2$DIII]Cl$_2$ directly with SiCl$_4$ in refluxing quinoline.

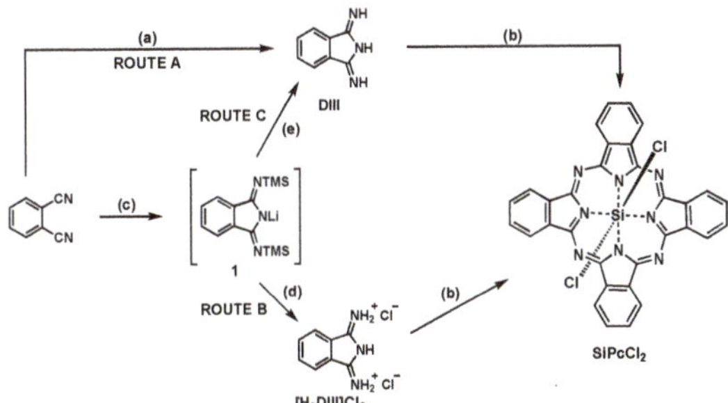

Scheme 1. Synthetic routes in the preparation of SiPcCl$_2$. Reagents and conditions: (**a**) (i) Na, MeOH, (ii) NH$_{3(g)}$, Δ; (**b**) SiCl$_4$, Quinoline, Δ; (**c**) LiN(TMS)$_2$·Et$_2$O, Toluene; (**d**) HCl$_{(g)}$; (**e**) (i) TMSCl, (ii) silica gel.

Although SiPcCl$_2$ can be synthesized via Route B in 21% yield, employing gaseous hydrochloric acid to isolate [H$_2$DIII]Cl$_2$ does not eliminate the use of corrosive gases. Alternatively, quenching 1 with chlorotrimethylsilane (TMSCl) affords the silylated DIII intermediate, while the LiCl by-product precipitates out of solution (Route C). Protonation and purification can be achieved upon passing the resulting brown solution through silica gel, affording DIII as a pale orange solid. Subsequent treatment of the DIII with SiCl$_4$ in refluxing quinoline affords SiPcCl$_2$ as a purplish blue matte solid in 51% yield. Via Route C, we have therefore established a safer alternative to that previously published, as we have not only eliminated the necessity for highly reactive reagents and corrosive gases, but we have also ascertained a method in which quantifiable amounts of each reagent necessary can be used. Overall, Route C offers the best alternative in terms of yield and purity of the final product and, through the use of LiN(TMS)$_2$·Et$_2$O in place of sodium metal and ammonia gas, paves the way for the development of additional derivatives that would otherwise not be possible with the previous synthetic routes.

To test the efficacy of the various routes to prepare SiPcCl$_2$, and to ensure the quality of the materials is maintained, samples from routes A–C were functionalized at the axial position with 3,4,5-trifluorophenoxy substituents and implemented into OTFTs. In that regard, bis(3,4,5-trifluorophenoxy) silicon phthalocyanine ((345F)$_2$-SiPc) was synthesized from SiPcCl$_2$ according to the literature [28], employing SiPcCl$_2$ from Routes A, B and C. In all cases, the (345F)$_2$-SiPc was purified by train sublimation and the products were obtained in similar yields with similar purities.

UV–Vis absorption spectroscopy revealed no difference between all the (345F)$_2$-SiPc derivatives in this study (Figure S1) and those reported in the literature [28]. (345F)$_2$-SiPc has previously been reported to have a good thermal stability (decomposition temperature of >400 °C), significant pi–pi interaction by single crystal X-ray diffraction and a highest occupied molecular orbital (HOMO) level of −5.90 eV and a respective lowest unoccupied molecular orbital energy level of −4.05 eV (Figure 2c) [28]. Organic thin-film transistors (OTFTs) were therefore fabricated using (345F)$_2$-SiPc (from Routes A, B and C) by thermal evaporation in a bottom-contact, bottom-gate configuration. Regardless of the synthetic route for the preparation of (345F)$_2$-SiPc, an average electron field-effect mobility (μ_e) of 3.1×10^{-3} cm^2 V^{-1} s^{-1} and threshold voltage (V$_T$) of 25.6 V were obtained. Although modest, this performance is superior to previously reported silicon-based phthalocyanine OTFT devices fabricated under similar conditions using bis(benzoate) silicon phthalocyanine as the semiconductor, where an average μ_e of 4.9×10^{-4} cm^2 V^{-1} s^{-1} and V$_T$ of 25.9 V were obtained [22]. For (345F)$_2$-SiPc, performance varied over 75 devices with an average electron field-effect mobility of 3.1×10^{-3} cm^2 V^{-1} s^{-1} and threshold voltage of 25.6 V, an on/off current ratio (I$_{on/off}$) ranging between 10^3 and 10^5 and a maximum μ_e of 1.64×10^{-2} cm^2 V^{-1} s^{-1} for all synthetic routes. Characteristic output and transfer curves for (345F)$_2$-SiPc devices are shown in Figure 2. We have previously shown that when integrated into bottom-gate, bottom-contact OTFTs, (345F)$_2$-SiPc and other SiPc derivatives experienced significant variations in performance as a function of channel lengths due to substantial contact resistance (as apparent in Figure 2a) [29]. Current efforts involve device engineering, including electrode material selection as a way to reduce contact resistance and amplify n-type performance.

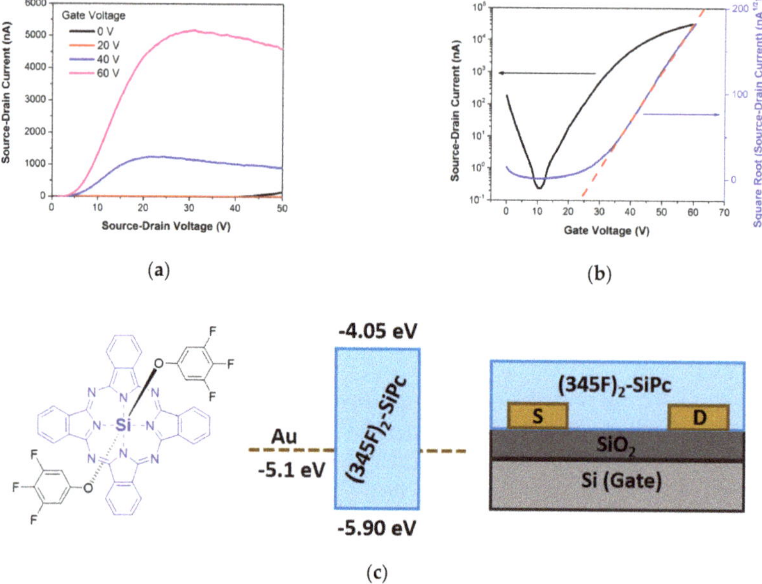

Figure 2. Characteristic output curve (**a**) and transfer curve (**b**) for bis(3,4,5-trifluorophenoxy) silicon phthalocyanine ((345F)$_2$-SiPc) bottom-gate, bottom-contact (BGBC) organic thin-film devices with L = 5 um. (**c**) is the chemical structure of ((345F)$_2$-SiPc), the respective highest occupied molecular orbital (HOMO) and lowest unoccupied molecular orbital (LUMO) levels of (345F)$_2$-SiPc and a diagram of the BGBC organic thin-film transistor (OTFT) structure (from left to right) [28].

3. Conclusions

In closing, we have demonstrated an efficient and effective method for the preparation of SiPcCl$_2$ and, upon axial substitution, have confirmed their n-type mobilities, which are the highest reported to date for SiPc-based OTFTs. Having developed the synthetic methodology to isolate SiPcCl$_2$ using less toxic, less corrosive, non-gaseous reagents, we can now begin to explore other derivatives with modified molecular frameworks. This therefore provides an avenue to control the energy levels of the frontier MOs, enabling the development of air stable n-type OSCs, in synergy with the combined morphological control through axial substitution. Such studies are currently underway.

4. Materials and Methods

4.1. General Methods and Procedures

The reagents phthalonitrile (TCI, Philadelphia, PA, USA), trimethylsilyl trifluoromethanesulfonate (Synquest, Alachua, FL, USA), chlorotrimethylsilane (Acros, Fair Lawn, NJ, USA), silicon tetrachloride (Sigma, Sheboygan, WI, USA) and 1,3-diiminoisoindoline (TCI, Philadelphia, PA, USA) were obtained commercially and used as received. Lithium bis(trimethylsilylamide) etherate [30], HCl gas [31] and bis(3,4,5-trifluorophenoxy) silicon phthalocyanine [32] were prepared as outlined in literature. All solvents were ACS grade; dry solvents were obtained by passing through activated alumina on a J.C. Meyer solvent purification system. Quinoline (Oakwood, Estill, SC, USA) was distilled over zinc dust. All reactions were performed under an atmosphere of dry nitrogen. NMR spectra were run in either DMSO-d$_6$ or MeCN-d$_3$ solutions at room temperature on a Bruker Avance 300 or 400 MHz spectrometer (Billerica, MA, USA, Figures S2–S4). ^{19}F-NMR spectra were referenced to trifluoroacetic acid at −76.55 ppm (Figure S5). IR spectra of solid samples were recorded on an Agilent Technologies Cary 630 FT-IR spectrometer (Cary, NC, USA).

4.1.1. Preparation of SiPcCl$_2$ (via Route A)

Reaction was performed analogous to previously reported literature starting with 1 g of commercial diiminoisoindoline [27]. Crude yield 0.76 g (1.2 mmol, 73%). Samples were characterized spectroscopically and compared to references in literature [33]. HR-MS (EI) for C$_{32}$H$_{16}$Cl$_2$N$_8$Si [M]$^+$: Calcd, 610.0644; Found, 610.0635.

4.1.2. Preparation of 2,3-Dihydro-1H-Isoindole-1,3-Bis(Iminium) Bis(Trifluoromethanesulfonate) [H$_2$DIII][OTf]$_2$ (via Route B)

A solution of lithium bis(trimethylsilyl)amide etherate (4.7 g, 19.5 mmol) in dry toluene (200 mL) was added dropwise to a stirring tolyl solution (150 mL) of phthalonitrile (2.5 g; 19.5 mmol) and subsequently stirred overnight. Hydrochloric acid was bubbled into the mixture for 25 minutes. The resulting green slurry was stirred for 30 min, followed by filtration in vacuo and washed thrice with MeCN to afford the green solid of [H$_2$DIII]Cl$_2$. IR ν_{max} = 3120 (br, w), 2672 (br, m), 2545 (br, m), 2347 (br, w), 1671 (s), 1601 (m), 1527 (m), 1475 (m), 1463 (m), 1353 (m), 1318 (m), 1276 (m), 1179 (m), 1086 (w), 1043 (s), 912 (w), 846 (s), 826 (s, br), 787 (s), 746 (m), 728 (m), 690 (s). This material is essentially insoluble in organic solvents; its identity was determined by metathesis to the more soluble triflate salt by the following procedure: Trimethylsilyl trifluoromthanesulfonate (10.6 mL, 58.5 mmol) was added to a slurry of [H$_2$DIII]Cl$_2$ in dry MeCN (25 mL) and stirred for 1 h. The green slurry was filtered in vacuo and washed thrice with DCE to afford a green solid of [H$_2$DIII][OTf]$_2$. Yield 4.8441 g (10.9 mmol, 56%). ^{1}H-NMR (δ, CD$_3$CN, RT, 300 MHz): 8.36 (m, 2H), 8.11 (m, 2H). ^{19}F-NMR (δ, CD$_3$CN, RT, 300 MHz): −79.20 (s, 6F). IR ν_{max} = 3252 (w), 3109 (w), 2967 (m, br), 1678 (m), 1606 (w), 1478 (w), 1467 (w), 1353 (w), 1300 (m), 1284 (m), 1271 (m), 1236 (m), 1217 (s), 1192 (s), 1167 (s), 1091 (w), 1077 (m), 1033 (s), 1022 (s), 978 (w), 868 (m), 840 (m), 783 (m), 764 (m), 748 (m), 697 (m).

4.1.3. Preparation of Silicon Phthalocyanine Dichloride (via Route B)

A solution of lithium bis(trimethylsilyl)amide etherate (4.7 g, 19.5 mmol) in dry toluene (200 mL) was added dropwise to a stirring solution of phthalonitrile (2.5 g; 19.5 mmol). After 5 h, the reaction mixture was cooled in an ice bath and hydrochloric acid was bubbled into the mixture for 25 min. The resulting green slurry was stirred for 30 min followed by filtration in vacuo and washed thrice with MeCN. To the filtered solid, dry quinoline (75 mL) and silicon tetrachloride (3.2 mL, 28.1 mmol) were added. The stirring slurry was heated to 180 °C for 4 h producing a visible red microcrystalline solid. The solid was filtered and washed with benzene, followed by methanol and acetone, respectively. Crude yield 0.63 g (1.0 mmol, 21%). Samples were characterized spectroscopically and compared to references in literature [33]. HR-MS (EI) for $C_{32}H_{16}Cl_2N_8Si$ $[M]^+$: Calcd, 610.0644; Found, 610.0632.

4.1.4. Preparation of DIII (via Route C)

A solution of lithium bis(trimethylsilyl)amide etherate (4.7 g, 19.5 mmol) in toluene (200 mL) was added dropwise to a stirring tolyl solution (150 mL) of phthalonitrile (2.5 g; 19.5 mmol) and subsequently stirred overnight. Chlorotrimethylsilane (2.2 mL, 19.5 mmol) was added to the solution and refluxed for an additional 24 h. Lithium chloride was removed through hot filtration of the mixture in vacuo. The brown filtrate was passed through a column of silica and washed with DCM and EtOAc, respectively. The desired product was collected with a 25% MeOH in EtOAc solution (v/v). The solvent of the filtrate was removed in vacuo and the resulting brown solid was washed with ether, DCM, and EtOAc consecutively, resulting in a light orange solid. Yield 1.23 g (8.5 mmol, 44%). Samples were characterized spectroscopically and compared to references in literature [34]. ^{1}H-NMR (δ, CD_3CN, RT, 400 MHz): 7.76 (m, 2H), 7.59 (m, 2H).

4.1.5. Preparation of Silicon Phthalocyanine Dichloride (via Route C)

Reaction was performed analogously to previously reported literature starting with 1 g of diiminoisoindoline (via Route C) [27]. Crude yield 0.54 g (0.9 mmol, 51%). Samples were characterized spectroscopically and compared to references in literature [33]. HR-MS (EI) for $C_{32}H_{16}Cl_2N_8Si$ $[M]^+$: Calcd, 610.0644; Found, 610.0637.

4.1.6. Preparation of Bis(3,4,5-trifluorophenoxy) Silicon Phthalocyanine

Reaction was performed analogous to previously reported literature using as prepared $SiPcCl_2$ from Routes A–C. Samples were characterized spectroscopically and compared to references in literature [29]. HR-MS (EI) for $C_{38}H_{18}N_8OF_3Si$ $[M–C_6H_2OF_3]^+$: Calcd, 687.1325; Found (Route A), 687.1311; Found (Route B), 687.1330; Found (Route C), 687.1315.

4.1.7. Electrical Testing

Bottom-gate, bottom-contact OTFTs made with $(345F)_2$-SiPc active layers were made starting with prefabricated substrates manufactured by Fraunhofer IPMS. These substrates had a doped silicon base/gate, a 230 nm SiO_2 dielectric, and pre-patterned source-drain electrodes made from gold with an indium tin oxide adhesion layer (W = 2000 μm, L = 2.5, 5, 10, 20 μm). Substrates were rinsed thoroughly with acetone to remove the protective resist and then treated for 15 min in oxygen plasma. Before transferring into a 1% v/v solution of octyltrichlorosilane (OTS, Sigma, 97%) in toluene, the substrates were rinsed with water and isopropanol, then dried with nitrogen. After 60 min heating at 70 °C, the substrates were removed from solution, rinsed with toluene and isopropanol, then dried under vacuum at 70 °C for 30–60 min. Dried, OTS-treated substrates were transferred into a vacuum chamber where $(345F)_2$-SiPc was evaporated onto their surface at a rate of 0.3Å/s until their thickness reached the target of 300Å as measured using a quartz crystal monitor. Finished devices were transferred in an evacuated capsule to the custom OTFT testing setup, oesProbe A10000-P290 (Element Instrumentation Inc. & Kreus Design Inc.), and tested under vacuum (P < 0.1 Pa) without having been exposed to air. Transfer characteristics were obtained in the range 0 V < V_{GS} < 60 V, increased step-wise every 100 ms, with V_{DS} set to 50 V. Field-effect mobility (μ_e) and threshold voltage (V_T) were calculated from the slope and x-intercept of Equation (1) in the range 35 V < V_{GS} < 45 V, with the dielectric capacitance

C_i calculated as $k\varepsilon_0/t$, where t is the dielectric thickness and k is the relative dielectric constant of SiO_2 (3.9).

$$I_{DS} = \frac{\mu C_i W}{2L}(V_{GS} - V_T) \tag{1}$$

Supplementary Materials: The following are available online at http://www.mdpi.com/1996-1944/12/8/1334/s1, Figure S1: Normalized absorbance spectra of $(345F)_2$-SiPc via Routes A (black), B (blue) and C (red) in toluene solutions, Figure S2: 1H-NMR spectra of DIII from Route B at 400 MHz in MeCN-d$_3$. Spectrum referenced to solvent residual peak at 1.94 ppm, Figure S3: ^{1}H-NMR spectra of [H$_2$DIII][OTf]$_2$ at 300 MHz in MeCN-d$_3$. Spectrum referenced to solvent residual peak at 1.94 ppm, Figure S4: ^{1}H-NMR spectra of DIII from commercial source (blue) and from Route B (red) in DMSO-d$_6$ at 400 MHz. Spectra referenced to residual solvent peak at 2.50 ppm, Figure S5. ^{19}F-NMR spectra of [H$_2$DIII][OTf]$_2$ at 300 MHz in MeCN-d$_3$. Spectrum referenced to F$_3$CCOH at −76.55 ppm.

Author Contributions: All authors contributed equally to this work.

Funding: The authors thank the University of Ottawa, the Canadian Foundation for Innovation (CFI), NSERC (Discovery grants and PGSD programs) and the Ontario Research Fund. N.J.Y. thanks NSERC for the Canada Graduate Scholarship (CGS D), O.A.M and T.M.G. are also grateful for the Ontario graduate student scholarship (OGS).

Conflicts of Interest: The authors declare no conflict of interest.

References

1. Miao, Q. N-Heteropentacenes and N-Heteropentacenequinones: From molecules to semiconductors. *Synlett* **2012**, *23*, 326–336. [CrossRef]

2. Quinn, J.T.E.; Zhu, J.X.; Li, X.; Wang, J.L.; Li, Y.N. Recent progress in the development of n-type organic semiconductors for organic field effect transistors. *J. Mat. Chem. C* **2017**, *5*, 8654–8681. [CrossRef]

3. Leznoff, C.C.; Lever, A.B.P. *Phthalocyanines: Properties and applications*; Wiley: Hoboken, NJ, USA, 1996; Volume 4, pp. 1–536.

4. Lu, H.; Kobayashi, N. Optically active porphyrin and phthalocyanine systems. *Chem. Rev.* **2016**, *116*, 6184–6261. [CrossRef] [PubMed]

5. Dahlen, M.A. The phthalocyanines - A new class of synthetic pigments, and dyes. *Ind. Eng. Chem.* **1939**, *31*, 839–847. [CrossRef]

6. Sorokin, A.B. Phthalocyanine metal complexes in catalysis. *Chem. Rev.* **2013**, *113*, 8152–8191. [CrossRef]

7. Bian, Y.Z.; Jiang, J.Z. Recent advances in phthalocyanine-based functional molecular materials. In *50 Years of Structure and Bonding—The Anniversary Volume*; Mingos, M.P., Ed.; Springer: Cham, Switzerland, 2015; Volume 172, pp. 159–199.

8. Ishikawa, N. Phthalocyanine-based magnets. In *Functional Phthalocyanine Molecular Materials*; Springer: Berlin, Germany, 2010; Volume 135, pp. 211–228.

9. Melville, O.A.; Lessard, B.H.; Bender, T.P. Phthalocyanine-based organic thin-film transistors: A review of recent advances. *ACS Appl. Mater. Interfaces* **2015**, *7*, 13105–13118. [CrossRef]

10. de la Torre, G.; Vazquez, P.; Agullo-Lopez, F.; Torres, T. Phthalocyanines and related compounds: organic targets for nonlinear optical applications. *J. Mater. Chem.* **1998**, *8*, 1671–1683. [CrossRef]

11. Shao, X.N.; Wang, S.R.; Li, X.G.; Su, Z.J.; Chen, Y.; Xiao, Y. Single component p-, ambipolar and n-type OTFTs based on fluorinated copper phthalocyanines. *Dyes Pigments* **2016**, *132*, 378–386. [CrossRef]

12. Song, D.; Zhu, F.; Yu, B.; Huang, L.H.; Geng, Y.H.; Yana, D.H. Tin (IV) phthalocyanine oxide: An air-stable semiconductor with high electron mobility. *Appl. Phys. Lett.* **2008**, *92*. [CrossRef]

13. Song, D.; Wang, H.B.; Zhu, F.; Yang, J.L.; Tian, H.K.; Geng, Y.H.; Yan, D.H. Phthalocyanato tin(IV) dichloride: An air-stable, high-performance, n-type organic semiconductor with a high field-effect electron mobility. *Adv. Mater.* **2008**, *20*, 2142–2144. [CrossRef]

14. Matumoto, A.; Hoshino, N.; Akutagawa, T.; Matsuda, M. N-Type Semiconducting Behavior of Copper Octafluorophthalocyanine in an Organic Field-Effect Transistor. *Appl. Sci.* **2017**, *7*, 1111. [CrossRef]

15. Zysman-Colman, E.; Ghosh, S.S.; Xie, G.H.; Varghese, S.; Chowdhury, M.; Sharma, N.; Cordes, D.B.; Slawin, A.M.Z.; Samuel, I.D.W. Solution-processable silicon phthalocyanines in electroluminescent and photovoltaic devices. *ACS Appl. Mater. Interfaces* **2016**, *8*, 9247–9253. [CrossRef]

16. Pearson, A.J.; Plint, T.; Jones, S.T.E.; Lessard, B.H.; Credgington, D.; Bender, T.P.; Greenham, N.C. Silicon phthalocyanines as dopant red emitters for efficient solution processed OLEDs. *J. Mater. Chem. C* **2017**, *5*, 12688–12698. [CrossRef]

17. Plint, T.; Lessard, B.H.; Bender, T.P. Assessing the potential of group 13 and 14 metal/metalloid phthalocyanines as hole transport layers in organic light emitting diodes. *J. Appl. Phys.* **2016**, *119*. [CrossRef]

18. Lessard, B.H.; Dang, J.D.; Grant, T.M.; Gao, D.; Seferos, D.S.; Bender, T.P. Bis(tri-n-hexylsilyl oxide) silicon phthalocyanine: A unique additive in ternary bulk heterojunction organic photovoltaic devices. *ACS Appl. Mater. Interfaces* **2014**, *6*, 15040–15051. [CrossRef]

19. Lessard, B.H.; White, R.T.; Al-Amar, M.; Plint, T.; Castrucci, J.S.; Josey, D.S.; Lu, Z.-H.; Bender, T.P. Assessing the potential roles of silicon and germanium phthalocyanines in planar heterojunction organic photovoltaic devices and how pentafluoro phenoxylation can enhance pi-pi interactions and device performance. *ACS Appl. Mater. Interfaces* **2015**, *7*, 5076–5088. [CrossRef] [PubMed]

20. Dang, M.-T.; Grant, T.M.; Yan, H.; Seferos, D.S.; Lessard, B.H.; Bender, T.P. Bis(tri-n-alkylsilyl oxide) silicon phthalocyanines: a start to establishing a structure property relationship as both ternary additives and non-fullerene electron acceptors in bulk heterojunction organic photovoltaic devices. *J. Mater. Chem. A* **2017**, *5*, 12168–12182. [CrossRef]

21. Grant, T.M.; Gorisse, T.; Dautel, O.; Wantz, G.; Lessard, B.H. Multifunctional ternary additive in bulk heterojunction OPV: increased device performance and stability. *J. Mater. Chem. A* **2017**, *5*, 1581–1587. [CrossRef]

22. Honda, S.; Ohkita, H.; Benten, H.; Ito, S. Selective dye loading at the heterojunction in polymer/fullerene solar cells. *Adv. Energy Mater.* **2011**, *1*, 588–598. [CrossRef]

23. Melville, O.A.; Grant, T.M.; Lessard, B.H. Silicon phthalocyanines as N-type semiconductors in organic thin film transistors. *J. Mater. Chem. C* **2018**, *6*, 5482–5488. [CrossRef]

24. Joyner, R.D.; Cekada, J.; Linck, R.G.; Kenney, M.E. Diphenoxysilicon Phthalocyanine. *J. Inorg. Nucl. Chem.* **1960**, *15*, 387–388. [CrossRef]

25. Joyner, R.D.; Kenney, M.E. Phthalocyaninosilicon Compounds. *Inorg. Chem.* **1962**, *1*, 236–238. [CrossRef]

26. Barrett, P.A.; Dent, C.E.; Linstead, R.P. Phthalocyanines Part VII Phthalocyanine as a co-ordinating group-A general investigation of the metallic derivatives. *J. Chem. Soc.* **1936**, 1719–1736. [CrossRef]

27. Lowery, M.K.; Starshak, A.J.; Esposito, J.N.; Krueger, P.C.; Kenney, M.E. Dichloro(phthalocyanino)silicon. *Inorg. Chem.* **1965**, *4*, 128. [CrossRef]

28. Ishidai, K.; Ookubo, K. Production Process for Colorant, Colorant Composition, Toner, Ink for Ink Jet Recording and Color Filter. European Patent EP2644660A1, 2 October 2013.

29. Melville, O.A.; Grant, T.M.; Mirka, B.; Boileau, N.T.; Park, J.; Lessard, B.H. Ambipolarity and air stability of silicon phthalocyanine organic thin-film transistors. *Adv. Electron. Mater.* **2019**, in press. [CrossRef]

30. Boeré, R.T.; Oakley, R.T.; Reed, R.W. Preparation of *N,N,N'*-tris(trimethylsilyl)amidines; a convenient route to unsubstituted amidines. *J. Organomet. Chem.* **1987**, *331*, 161–167. [CrossRef]

31. Arnáiz, F.J. A convenient way to generate hydrogen chloride in the freshman lab. *J. Chem. Educ.* **1995**, *72*, 1139. [CrossRef]

32. Lessard, B.H.; Grant, T.M.; White, R.; Thibau, E.; Lu, Z.-H.; Bender, T.P. The position and frequency of fluorine atoms changes the electron donor/acceptor properties of fluorophenoxy silicon phthalocyanines within organic photovoltaic devices. *J. Mater. Chem. A* **2015**, *3*, 24512–24524. [CrossRef]

33. Park, H.S.; Rya, H.; Jang, N.H. Synthesis and characterization of phthalocyaninatosilicon with bridging ligands (L) (L = dimethylsilane, diphenylsilane, methylphenylsilane). *Bull. Korean Chem. Soc.* **2016**, *37*, 207–212. [CrossRef]

34. Elvidge, J.A.; Golden, J.H. Conjugated macrocycles. Part XXVIII.* Adducts from diimimisoindoline and arylene-m-diamines, and a new type of crossconjugated macrocycle with three-quarters of the chromophore of phthalocyanine. *J. Chem. Soc.* **1957**. [CrossRef]

Article

Push-Pull Chromophores Based on the Naphthalene Scaffold: Potential Candidates for Optoelectronic Applications

Corentin Pigot [1,*], Guillaume Noirbent [1,*], Thanh-Tuân Bui [2], Sébastien Péralta [2], Didier Gigmes [1], Malek Nechab [1] and Frédéric Dumur [1,*]

[1]   Aix Marseille Univ, CNRS, ICR UMR7273, F-13397 Marseille, France; didier.gigmes@univ-amu.fr (D.G.); malek.nechab@univ-amu.fr (M.N.)

[2]   Laboratoire de Physicochimie des Polymères et des Interfaces (LPPI), Université de Cergy Pontoise, 5 mail Gay Lussac, F-95000 Neuville-sur-Oise, France; tbui@u-cergy.fr (T.-T.B.); sebastien.Peralta@u-cergy.fr (S.P.)

*   Correspondence: corentin.pigot@univ-amu.fr (C.P.); Guillaume.noirbent@outlook.fr (G.N.); frederic.dumur@univ-amu.fr (F.D.)

Received: 5 April 2019; Accepted: 22 April 2019; Published: 24 April 2019

Abstract: A series of ten push-pull chromophores comprising 1*H*-cyclopenta[*b*]naphthalene-1,3(2*H*)-dione as the electron-withdrawing group have been designed, synthesized, and characterized by UV-visible absorption and fluorescence spectroscopy, cyclic voltammetry and theoretical calculations. The solvatochromic behavior of the different dyes has been examined in 23 solvents and a positive solvatochromism has been found for all dyes using the Kamlet-Taft solvatochromic relationship, demonstrating the polar form to be stabilized in polar solvents. To establish the interest of this polyaromatic electron acceptor only synthesizable in a multistep procedure, a comparison with the analog series based on the benchmark indane-1,3-dione (1*H*-indene-1,3(2*H*)-dione) has been done. A significant red-shift of the intramolecular charge transfer band has been found for all dyes, at a comparable electron-donating group. Parallel to the examination of the photophysical properties of the different chromophores, a major improvement of the synthetic procedure giving access to 1*H*-cyclopenta[*b*]naphthalene-1,3(2*H*)-dione has been achieved.

Keywords: push-pull dyes; chromophore; naphthalene; solvatochromism

---

1. Introduction

During the past decades, push-pull molecules have attracted much attention due to their numerous applications ranging from non-linear optical (NLO) applications [1,2] to organic photovoltaics (OPVs) [3,4], organic field effect transistors (OFETs) [5], organic light-emitting diodes (OLEDs) [6], photorefractive applications [7], colorimetric pH sensors [8], ions detection [9], biosensors [10], gas sensors [11], or photoinitiators of polymerization [12–16]. Typically, push-pull chromophores are based on an electron donor and an electron acceptor connected to each other by mean of a saturated or a conjugated spacer [17]. As the first manifestation of the mutual interaction between the two partners, a broad absorption band corresponding to the intramolecular charge transfer (ICT) interaction can be detected in the visible to the near-infrared region. Position of this absorption band typically depends on the electron-donating ability of the donor and the electron-releasing ability of the acceptor, and a bathochromic shift of this transition is observed upon improvement of the strength of the donor and/or the acceptor. This ICT band can even be detected in the near and far infrared region for electron-acceptors based on poly(nitrofluorenes) [18]. Among electron acceptors, indane-1,3-dione (1*H*-indene-1,3(2*H*)-dione) **EA1** has been extensively studied due to its commercial availability, its low cost and the possibility to design push-pull dyes by a mean of one of the simplest reaction of Organic Chemistry, namely the Knoevenagel reaction [19–23]. By chemical engineering, its electron-withdrawing ability can be drastically improved by condensation of one or two malononitrile units under basic conditions, furnishing 2-(3-oxo-2,3-dihydro-1*H*-inden-1-ylidene)malononitrile **EA2** [24] and 2,2′-(1*H*-indene-1,3 (2*H*)-diylidene)dimalononitrile **EA3** [25] (see Figure 1). Based on the strong electron-withdrawing abilities of **EA2** and **EA3**, push-pull dyes with non-fullerene acceptors and small energy gaps have been developed [26,27]. As a possible alternative to improve the electron-accepting ability, extension of the aromaticity of the acceptor can be envisioned. Using this strategy, a red-shift of the ICT band combined with an enhancement of the molar extinction coefficient can be both obtained for these new push-pull derivatives, relative to that obtained with the parent electron acceptor.

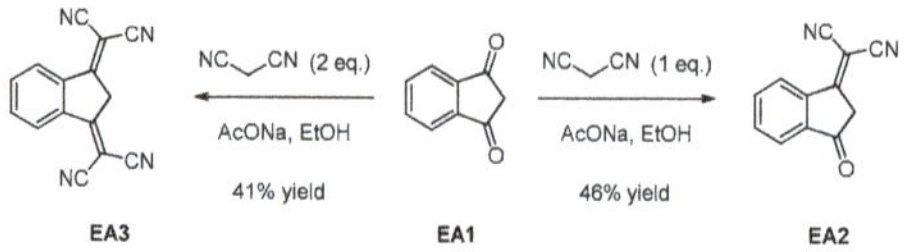

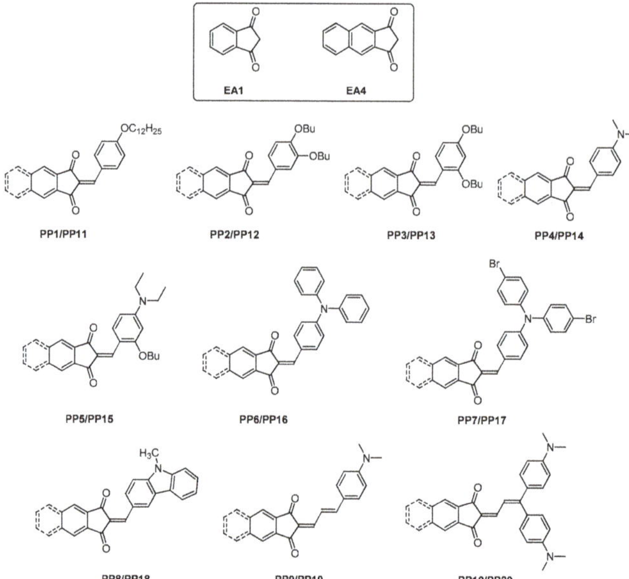

Figure 1. Electron acceptors **EA1–EA3**.

In this article, an extended version of the well-known indane-1,3-dione, i.e., 1*H*-cyclopenta[*b*] naphthalene-1,3(2*H*)-dione **EA4**, has been used for the design of ten push-pull dyes. It has to be noticed that even if the synthesis of **EA4** is reported in the literature since 2006 [28], only one report mentions the development of anion sensors with **EA4** in 2013 [29] and the first push-pull dyes have been developed in 2017 for photovoltaics applications [30,31]. Considering the scarcity of studies devoted to this electron acceptor, a series of 10 push-pull chromophores **PP1–PP10** have been developed with **EA4**. To evidence the contribution of the additional aromatic ring in **EA4** relative to that of **EA1**, 10 dyes **PP11–PP20** comprising **EA1** as the electron acceptor have been synthesized for comparison (see Figure 2).

Figure 2. Chemical structures of the 20 dyes **PP1–PP20** examined in this study. **PP1–PP10** were made from **EA4**, while **PP11–PP20** were made from **EA1**.

The photophysical properties of the series of 20 dyes as well as their electrochemical properties have been investigated. To support the experimental results, theoretical calculations have been carried out. Finally, the solvatochromic properties have also been examined.

2. Materials and Methods

All reagents and solvents were purchased from Aldrich, Alfa Aesar or TCI Europe and used as received without further purification. Mass spectroscopy was performed by the Spectropole of Aix-Marseille University. Electron spray ionization (ESI) mass spectral analyses were recorded with a 3200 QTRAP (Applied Biosystems SCIEX) mass spectrometer. The HRMS mass spectral analysis was performed with a QStar Elite (Applied Biosystems SCIEX) mass spectrometer. Elemental analyses were recorded with a Thermo Finnigan EA 1112 elemental analysis apparatus driven by the Eager 300 software. ^{1}H and ^{13}C nuclear magnetic resonance (NMR) spectra were determined at room temperature in 5 mm outer diameter (o.d.) tubes on a Bruker Avance 400 spectrometer of the Spectropole: ^{1}H (400 MHz) and ^{13}C (100 MHz). The ^{1}H chemical shifts were referenced to the solvent peak CDCl$_3$ (7.26 ppm) and the ^{13}C chemical shifts were referenced to the solvent peak CDCl$_3$ (77 ppm). UV-visible absorption spectra were recorded on a Varian Cary 50 Scan UV Visible Spectrophotometer, with concentration of 5×10^{-3} M, corresponding to diluted solutions. Fluorescence spectra were recorded using a Jasco FP 6200 spectrometer. The electrochemical properties of the investigated compounds were measured in acetonitrile by cyclic voltammetry, scan rate 100 mV·s^{-1}, with tetrabutylammonium tetrafluoroborate (0.1 M) as a supporting electrolyte in a standard one-compartment, three-electrode electrochemical cell under an argon stream using a VSP BioLogic potentiostat. The working, pseudo-reference and counter electrodes were platinum disk ($\varnothing$ = 1 mm), Ag wire, and Au wire gauze, respectively. Ferrocene was used as an internal standard, and the potentials are referred to the reversible formal potential of this compound. Computational details: All quantum mechanical calculations were computed using Gaussian Package [32]. All geometry optimizations were performed using the density functional theory (DFT) with the global hybrid exchange-correlation functional B3LYP [33] and all minima on the potential energy surface were verified via a calculation of vibrational frequencies, ensuring no imaginary frequencies were present. The Pople double-zeta basis set with a double set of polarization functions on non-hydrogen atoms (6-3111G(d,p))[34,35] was used throughout. This computational approach was chosen in consistency with previous works, as it provides good agreement with experimental data. Excited states were probed using the time dependent density functional theory (TD-DFT) using the same function. All transitions (singlet-singlet) were calculated vertically with respect to the singlet ground state geometry. Solvent effects were taken into account by using the implicit polarizable continuum model (PCM) [36,37]. Dichloromethane (DCM) where chosen in analogy with the experiments. Computed spectra were simulated by convoluting each transition with Gaussians functions-centered on each absorption maximum using a constant full width at half maximum (FWHM) value of 0.2 eV. The assignment of electronic transitions for λ_{max} has been determined with GaussSum 3.0 software [38]. 4-(Dodecyloxy)benzaldehyde **D1** [39], 3,4-dibutoxy-benzaldehyde **D2** [40], 2,4-dibutoxy-benzaldehyde **D3** [41], 4-(diphenylamino)benzaldehyde **D6** [42], 4-(*bis*(4-bromophenyl)-amino)benzaldehyde **D7** [43], and 9-methyl-9*H*-carbazole-3-carbaldehyde **D8** [44], 3-(4-(dimethylamino)phenyl)acrylaldehyde **D9** [45], and 3,3-*bis*(4-(dimethylamino)phenyl)-acrylaldehyde **D10** [46] were synthesized, as previously reported, without modifications and in similar yields.

2.1. Synthesis of the Dyes

2.1.1. 2-Butoxy-4-(Diethylamino)Benzaldehyde **D5**

A mixture of 4-(diethylamino)salicylaldehyde (26.5 g, 137 mmol, M = 193.24 g/mol), 1-bromobutane (21.8 g, 159 mmol) and potassium carbonate (18.9 g, 152.2 mmol) were heated in *N,N*-dimethylformamide (DMF) (250 mL) at reflux overnight. The solution was concentrated before addition of water. Extraction of the product was carried out with diethyl ether. The ethereal solution was washed with several portions of water to remove remaining DMF, dried over $MgSO_4$ and concentrated to yield the product as an orange oil (32.1 g, 128.73 mmol, 94% yield). ^{1}H nmR (CDCl$_3$) δ: 0.91 (t, 3*H*, J = 7.4 Hz), 1.14 (t, 6*H*, J = 7.1 Hz), 1.42–1.48 (m, 2*H*), 1.71–1.78 (m, 2*H*), 3.34 (q, 4*H*, J = 7.1 Hz), 3.96 (t, 2*H*, J = 6.3 Hz), 5.95 (d, 1*H*, J = 2.3 Hz), 6.19 (dd, 1*H*, J = 9.0 Hz, J = 2.3 Hz), 7.63 (d, 1*H*, J = 9.0 Hz), 10.1 (s, 1*H*, CHO); ^{13}C nmR (CDCl$_3$) δ: 12.6, 13.8, 19.3, 31.2, 44.7, 67.8, 93.2, 104.2, 114.4, 130.1, 153.8, 163.9, and 187.1; HRMS (ESI MS) m/z: theor: 249.1729 found: 249.1733 ([M]$^{+\cdot}$ detected).

2.1.2. 1*H*-Cyclopenta[*b*]Naphthalene-1,3(2*H*)-Dione **EA4**

Diethyl naphthalene-2,3-dicarboxylate (10 g, 38.5 mmol, 1 eq., M = 272.30 g/mol) was suspended in extra dry EtOAc (24 mL) and NaH 95% in oil (2.44 g, 96.4 mmol, 2.5 eq.) was added. The reaction media was refluxed at 105 °C for 5 h. After cooling, the yellow solid was filtered off and thoroughly washed with a mixture of EtOH-Et$_2$O 50/50. Treatment of this solid with 200 mL of a 1 M HCl solution under reflux for 1.5 h furnished a new solid. After cooling, the solid was filtered off, washed with water and recrystallized in toluene (200 mL). The product was obtained as a brown solid (6.9 g, 35.27 mmol, 91% yield). ^{1}H nmR (CDCl$_3$) δ: 3.38 (s, 2*H*), 7.66–7.75 (m, 2*H*), 8.10–8.13 (m, 2*H*), 8.52 (s, 2*H*); ^{13}C nmR (CDCl$_3$) δ: 46.7, 124.3, 129.7, 130.6, 136.4, 138.2, and 197.6; HRMS (ESI MS) m/z: theor: 196.0524 found: 196.0526 ([M]$^{+\cdot}$ detected).

2.2. General Procedure for the Synthesis of **PP1–PP10**:

1*H*-Cyclopenta[b]naphthalene-1,3(2*H*)-dione **EA4** (0.5 g, 2.55 mmol) and the appropriate substituted benzaldehyde (2.55 mmol, 1. eq.) were dissolved in absolute ethanol (50 mL) and a few drops of piperidine were added. The reaction mixture was refluxed and progress of the reaction was followed by thin layer chromatography (TLC). After cooling, a precipitate formed. It was filtered off, washed several times with ethanol and dried under vacuum.

2.2.1. 2-(4-(Dodecyloxy)Benzylidene)-1*H*-Cyclopenta[b]Naphthalene-1,3(2*H*)-Dione **PP1**

4-(Dodecyloxy)benzaldehyde (0.74 g, 2.55 mmol), m$_{exp}$ = 1.05 g, 2.24 mmol, 88% yield. ^{1}H nmR (CDCl$_3$) δ: 0.88 (t, 3*H*, J = 6.2 Hz), 1.27–1.40 (m, 16*H*), 1.46 (qt, 2*H*, J = 7.5 Hz), 1.83 (qt, 2*H*, J = 6.7 Hz), 4.09 (t, 2*H*, J = 6.5 Hz), 7.02 (d, 2*H*, J = 8.7 Hz), 7.67–7.69 (m, 2*H*), 7.94 (s, 1*H*), 8.09–8.10 (m, 2*H*), 8.49 (d, 1*H*, J = 3.0 Hz), 8.62 (d, 2*H*, J = 8.7 Hz); ^{13}C nmR (CDCl$_3$) δ: 14.1, 22.7, 26.0, 29.1, 29.3, 29.55, 29.59, 29.63, 29.66, 31.92, 114.9, 123.8, 123.9, 126.5, 128.2, 129.0, 129.1, 130.4, 130.5, 135.6, 136.3, 136.5, 137.6, 137.7, 148.0, 164.2, 189.4, and 190.7; HRMS (ESI MS) m/z: theor: 468.2664 found: 468.2665 ([M]$^{+\cdot}$ detected).

2.2.2. 2-(3,4-Dibutoxybenzylidene)-1*H*-Cyclopenta[*b*]Naphthalene-1,3(2*H*)-Dione **PP2**

3,4-Dibutoxybenzaldehyde (0.64 g, 2.55 mmol), m$_{exp}$ = 0.92 g, 2.15 mmol, 84% yield. ^{1}H nmR (CDCl$_3$) δ: 1.01 (t, 3*H*, J = 7.6 Hz), 1.04 (t, 3*H*, J = 7.5 Hz), 1.50–1.64 (m, 4*H*), 1.82–1.94 (m, 4*H*), 4.14 (t, 2*H*, J = 6.5 Hz), 4.26 (t, 2*H*, J = 6.4 Hz), 6.97 (d, 1*H*, J = 8.4 Hz), 7.66–7.69 (m, 2*H*), 7.85 (d, 1*H*, J = 7.1 Hz), 7.92 (s, 1*H*), 8.07–8.10 (m, 2*H*), 8.49 (s, 2*H*), 8.83 (s, 1*H*); ^{13}C nmR (CDCl$_3$) δ: 13.8, 13.9, 19.2, 19.3, 31.0, 31.2, 68.8, 68.9, 112.1, 117.8, 123.8, 123.9, 126.9, 128.0, 129.0, 129.1, 130.4, 130.5, 131.7, 135.6, 136.3, 136.5, 137.6, 148.7, 148.8, 154.7, 189.5, and 190.7; HRMS (ESI MS) m/z: theor: 428.1988 found: 428.1987 ([M]$^{+\cdot}$ detected).

2.2.3. 2-(2,4-Dibutoxybenzylidene)-1*H*-Cyclopenta[*b*]Naphthalene-1,3(2*H*)-Dione **PP3**

2,4-Dibutoxybenzaldehyde (0.64 g, 2.55 mmol), m$_{exp}$ = 0.95 g, 2.22 mmol, 87% yield. ^{1}H nmR (CDCl$_3$) δ: 1.00 (t, 3*H*, J = 7.4 Hz), 1.04 (t, 3*H*, J = 7.4 Hz), 1.49–1.63 (m, 4*H*), 1.82 (qt, 2*H*, J = 7.7 Hz), 1.91 (qt, 2*H*, J = 6.4 Hz), 4.09 (t, 4*H*, J = 6.2 Hz), 6.44 (d, 1*H*, J = 0.9 Hz), 6.64 (dd, 1*H*, J = 11.0 Hz, J = 0.9 Hz), 7.64–7.67 (m, 2*H*), 8.06–8.09 (m, 2*H*), 8.44 (s, 2*H*), 8.62 (s, 1*H*), 9.37 (d, 1*H*, J = 9.0 Hz); ^{13}C nmR (CDCl$_3$) δ: 13.77, 13.83, 19.17, 19.32, 31.04-31.13, 68.2, 68.7, 98.8, 106.2, 116.5, 123.44, 123.48, 127.1, 128.7,

128.8, 130.3, 130.4, 135.8, 136.3, 136.4, 136.9, 137.7, 142.4, 163.1, 166.6, 189.5, and 191.0; HRMS (ESI MS) m/z: theor: 428.1988 found: 428.1986 ([M]$^{+\cdot}$ detected).

2.2.4. 2-(4-(Dimethylamino)Benzylidene)-1*H*-Cyclopenta[*b*]Naphthalene-1,3(2*H*)-Dione **PP4**

4-Dimethylaminobenzaldehyde (0.38 g, 2.55 mmol), m_{exp} = 0.62 g, 1.89 mmol, 74% yield. ^{1}H nmR (CDCl$_3$) δ: 3.16 (s, 6*H*), 6.74 (d, 2*H*, J = 9.0 Hz), 7.62–7.64 (m, 2*H*), 7.87 (s, 1*H*), 8.05–8.06 (m, 2*H*), 8.39 (d, 2*H*, J = 2.6 Hz), 8.61 (d, 2*H*, J = 8.6 Hz); ^{13}C nmR (CDCl$_3$) δ: 40.1, 111.5, 122.4, 122.9, 123.1, 125.1, 128.5, 128.6, 130.2, 130.3, 135.9, 136.1, 136.3, 137.8, 138.5, 148.5, 154.3, 189.6, and 191.4; HRMS (ESI MS) m/z: theor: 327.1259 found: 327.1260 ([M]$^{+\cdot}$ detected).

2.2.5. 2-(2-Butoxy-4-(Diethylamino)Benzylidene)-1*H*-Cyclopenta[b]Naphthalene-1,3(2*H*)-Dione **PP5**

2-Butoxy-4-(diethylamino)benzaldehyde (0.63 g, 2.55 mmol), m_{exp} = 1.02 g, 2.38 mmol, 94% yield. ^{1}H nmR (CDCl$_3$) δ: 1.04 (t, 3*H*, J = 7.4 Hz), 1.28 (t, 6*H*, J = 7.1 Hz), 1.60 (qt, 2*H*, J = 7.4 Hz), 1.92 (qt, 2*H*, J = 7.8 Hz), 3.50 (q, 4*H*, J = 7.1 Hz), 4.08 (t, 2*H*, J = 6.4 Hz), 6.03 (d, 1*H*, J = 2.0 Hz), 6.43 (dd, 1*H*, J = 9.3 Hz, J = 2.0 Hz), 7.59–7.62 (m, 2*H*), 8.03–8.05 (m, 2*H*), 8.34 (s, 2*H*), 8.58 (s, 1*H*), 9.50 (d, 1*H*, J = 9.3 Hz); ^{13}C nmR (CDCl$_3$) δ: 12.8, 13.9, 19.4, 31.1, 45.1, 68.2, 93.1, 105.1, 113.2, 122.37, 122.40, 123.1, 128.1, 128.2, 130.1, 130.2, 136.08, 136.11, 136.3, 137.8, 138.0, 155.2, 164.1, 189.8, and 191.9; HRMS (ESI MS) m/z: theor: 427.2147 found: 427.2149 ([M]$^{+\cdot}$ detected).

2.2.6. 2-(4-(Diphenylamino)Benzylidene)-1*H*-Cyclopenta[*b*]Naphthalene-1,3(2*H*)-Dione **PP6**

4-(Diphenylamino)benzaldehyde (0.70 g, 2.55 mmol), m_{exp} = 1.02 g, 2.26 mmol, 89% yield. ^{1}H nmR (CDCl$_3$) δ: 7.02 (d, 2*H*, J = 8.9 Hz), 7.20–7.24 (m, 6*H*), 7.38 (t, 4*H*, J = 8.2 Hz), 7.65–7.67 (m, 2*H*), 7.88 (s, 1*H*), 8.07–8.09 (m, 2*H*), 8.44 (d, 2*H*, J = 7.7 Hz), 8.50 (d, 2*H*, J = 8.9 Hz); ^{13}C nmR (CDCl$_3$) δ: 118.7, 123.50, 123.56, 125.7, 126.0, 126.7, 127.2, 128.8, 128.9, 129.8, 130.3, 130.4, 135.8, 136.3, 136.4, 137.3, 137.7, 145.6, 147.6, 153.1, 189.4, and 191.0; HRMS (ESI MS) m/z: theor: 451.1572 found: 451.1574 ([M]$^{+\cdot}$ detected).

2.2.7. 2-(2-Butoxy-4-(Diethylamino)Benzylidene)-1*H*-Cyclopenta[*b*]Naphthalene-1,3(2*H*)-Dione **PP7**

4-(*Bis*(4-bromophenyl)amino)benzaldehyde (1.1 g, 2.55 mmol, M = 431.13 g/mol), m_{exp} = 1.42 g, 2.33 mmol, 92% yield. ^{1}H nmR (CDCl$_3$) δ: 7.03–7.08 (m, 6*H*), 7.48 (d, 4*H*, J = 8.7 Hz), 7.66–7.69 (m, 2*H*), 7.88 (s, 1*H*), 8.07–8.11 (m, 2*H*), 8.46–8.52 (m, 4*H*); ^{13}C nmR (CDCl$_3$) δ: 118.6, 119.8, 123.78, 123.82, 127.1, 127.7, 128.2, 128.98, 129.08, 133.0, 135.7, 136.4, 136.5, 137.0, 137.6, 144.7, 147.0, 151.9, 189.3, and 190.7; HRMS (ESI MS) m/z: theor: 606.9783 found: 606.9787 ([M]$^{+\cdot}$ detected).

2.2.8. 2-((9-Methyl-9H-Carbazol-3-Yl)Methylene)-1*H*-Cyclopenta[*b*]Naphthalene-1,3(2*H*)-Dione **PP8**

9-Methyl-9*H*-carbazole-3-carbaldehyde (0.53 g, 2.55 mmol), m_{exp} = 0.84 g, 2.17 mmol, 85% yield. ^{1}H nmR (CDCl$_3$) δ: 3.91 (s, 3*H*), 7.37 (t, 1*H*, J = 6.7 Hz), 7.43 (d, 1*H*, J = 7.8 Hz), 7.48 (d, 1*H*, J = 8.7 Hz), 7.53 (t, 1*H*, J = 7.3 Hz), 7.66–7.69 (m, 2*H*), 8.05–8.12 (m, 2*H*), 8.20 (s, 1*H*), 8.28 (d, 1*H*, J = 7.3 Hz), 8.48 (d, 2*H*, J = 7.6 Hz), 8.76 (d, 1*H*, J = 8.0 Hz); ^{13}C nmR (CDCl$_3$) δ: 29.4, 108.8, 109.2, 120.8, 121.0, 123.2, 123.66, 123.68, 123.7, 125.3, 126.8, 127.5, 128.8, 128.9, 129.0, 130.38, 130.40, 133.9, 135.7, 136.3, 136.5, 137.8, 141.7, 144.5, 149.9, 189.5, and 190.9; HRMS (ESI MS) m/z: theor: 387.1259 found: 387.1263 ([M]$^{+\cdot}$ detected).

2.2.9. 2-(3-(4-(Dimethylamino)Phenyl)Allylidene)-1*H*-Cyclopenta[*b*]Naphthalene-1,3(2*H*)-Dione **PP9**

3-(4-(Dimethylamino)phenyl)acrylaldehyde (0.45 g, 2.55 mmol), m_{exp} = 756 mg, 2.14 mmol, 84% yield. ^{1}H nmR (CDCl$_3$) δ: 3.10 (s, 6*H*), 6.70 (d, 2*H*, J = 8.4 Hz), 7.36 (d, 1*H*, J = 15.1 Hz), 7.61–7.75 (m, 4*H*), 7.73 (d, 1*H*, J = 12.3 Hz), 8.03–8.07 (m, 2*H*), 8.34–8.39 (m, 3*H*); ^{13}C nmR (CDCl$_3$) δ: 40.1, 111.9, 119.8, 122.9, 123.3, 123.8, 126.3, 128.66, 128.72, 130.30, 130.35, 131.7, 136.2, 136.3, 136.5, 137.6, 147.6, 152.8, 154.8, 190.6, and 190.9; HRMS (ESI MS) m/z: theor: 353.1416 found: 353.1414 ([M]$^{+\cdot}$ detected).

2.2.10. 2-(3,3-Bis(4-(Dimethylamino)Phenyl)Allylidene)-1*H*-Cyclopenta[*b*]Naphthalene-1,3(2*H*)-Dione **PP10**

3,3-*Bis*(4-(dimethylamino)phenyl)acrylaldehyde (0.75 g, 2.55 mmol), m_{exp} = 1.06 g, 2.24 mmol, 88% yield. ^{1}H nmR (CDCl$_3$) δ: 3.08 (s, 6*H*), 3.09 (s, 6*H*), 6.68 (d, 2*H*, J = 9.0 Hz), 6.78 (d, 2*H*, J = 8.7 Hz), 7.24 (d, 2*H*, J = 10.2 Hz), 7.50 (d, 2*H*, J = 8.9 Hz), 7.60–7.63 (m, 2*H*), 7.78 (d, 1*H*, J = 12.9 Hz), 8.01–8.06 (m, 2*H*), 8.33 (d, 2*H*, J = 5.2 Hz), 8.42 (d, 2*H*, J = 12.9 Hz); ^{13}C nmR (CDCl$_3$) δ: 40.1, 40.2, 111.4, 111.5, 119.4, 122.4, 122.7, 125.8, 125.9, 128.3, 128.4, 128.8, 130.2, 132.5, 133.7, 136.1, 136.3, 136.7, 137.8, 146.6, 151.9, 152.4, 167.0, 191.0, and 191.1; HRMS (ESI MS) m/z: theor: 472.2151 found: 472.2148 ([M]$^{+\cdot}$ detected).

2.3. General Procedure for the Synthesis of *PP11–PP20*

Indane-1,3-dione (1*H*-indene-1,3(2*H*)-dione) **EA1** (0.37 g, 2.55 mmol) and the appropriate substituted benzaldehyde (2.55 mmol, 1 eq.) were dissolved in absolute ethanol (50 mL) and a few drops of piperidine were added. The reaction mixture was refluxed and progress of the reaction was followed by TLC. After cooling, a precipitate formed in most of the case. This latter was filtered off, washed several times with ethanol and dried under vacuum. For several chromophores (**PP11**, **PP12**, **PP13** or **PP20**), a purification by column chromatography on SiO$_2$ was required.

2.3.1. 2-(4-(Dodecyloxy)Benzylidene)-1*H*-Indene-1,3(2*H*)-Dione **PP11**

4-(Dodecyloxy)benzaldehyde (0.74 g, 2.55 mmol), m_{exp} = 790 mg, 1.89 mmol, 74% yield. ^{1}H nmR (CDCl$_3$) δ: 0.88 (t, 3*H*, J = 6.4 Hz), 1.21–1.54 (m, 18*H*), 1.83 (qt, 2*H*, J = 7.8 Hz), 4.07 (t, 2*H*, J = 6.5 Hz), 7.01 (d, 2*H*, J = 8.9 Hz), 7.77–7.85 (m, 2*H*), 7.85 (s, 1*H*), 7.97–8.00 (m, 2*H*), 8.54 (d, 2*H*, J = 8.9 Hz); ^{13}C nmR (CDCl$_3$) δ: 14.1, 22.7, 26.0, 29.1, 29.3, 29.54, 29.58, 29.62, 29.64, 31.9, 68.5, 114.9, 123.0, 126.3, 126.4, 134.8, 135.0, 137.3, 140.0, 142.4, 147.0, 163.8, 189.6, and 190.9; HRMS (ESI MS) m/z: theor: 418.2508 found: 418.2503 ([M]$^{+\cdot}$ detected).

2.3.2. 2-(3,4-Dibutoxybenzylidene)-1*H*-Indene-1,3(2*H*)-Dione **PP12**

3,4-Dibutoxybenzaldehyde (0.64 g, 2.55 mmol), m_{exp} = 820 mg, 2.17 mmol, 85% yield. ^{1}H nmR (CDCl$_3$) δ: 1.00 (t, 3*H*, J = 7.5 Hz), 1.02 (t, 3*H*, J = 7.5 Hz), 1.48–1.63 (m, 4*H*), 1.83–1.93 (m, 4*H*), 4.12 (t, 2*H*, J = 6.6 Hz), 4.23 (t, 2*H*, J = 6.4 Hz), 6.95 (d, 1*H*, J = 8.5 Hz), 7.77–7.82 (m, 4*H*), 7.96–8.00 (m, 2*H*), 8.72 (d, 1*H*, J = 1.7 Hz); ^{13}C nmR (CDCl$_3$) δ: 13.8, 13.9, 19.2, 19.3, 31.0, 31.2, 68.8, 68.9, 112.1, 117.6, 123.0, 126.2, 126.6, 131.2, 134.7, 134.9, 140.0, 142.5, 147.6, 148.8, 154.4, 189.7, and 190.9; HRMS (ESI MS) m/z: theor: 378.1831 found: 378.1832 ([M]$^{+\cdot}$ detected).

2.3.3. 2-(2,4-Dibutoxybenzylidene)-1*H*-Indene-1,3(2*H*)-Dione **PP13**

2,4-Dibutoxybenzaldehyde (0.64 g, 2.55 mmol), m_{exp} = 723 mg, 1.91 mmol, 75% yield. ^{1}H nmR (CDCl$_3$) δ: 1.00 (t, 3*H*, J = 7.4 Hz), 1.02 (t, 3*H*, J = 7.3 Hz), 1.46–1.64 (m, 6*H*), 1.76–1.94 (m, 4*H*), 4.07 (t, 2*H*, J = 6.4 Hz), 4.08 (t, 2*H*, J = 6.4 Hz), 6.42 (d, 1*H*, J = 2.2 Hz), 6.61 (dd, 1*H*, J = 9.0 Hz, J = 2.2 Hz), 7.73–7.76 (m, 2*H*), 7.94–7.97 (m, 2*H*), 8.51 (s, 1*H*), 9.22 (d, 1*H*, J = 9.0 Hz); ^{13}C nmR (CDCl$_3$) δ: 13.7, 13.8, 19.2, 31.1, 31.2, 68.2, 68.6, 98.8, 106.1, 116.1, 122.7, 122.8, 125.2, 134.5, 124.6, 136.5, 140.0, 141.3, 142.3162.7, 166.1, 189.8, and 191.2; HRMS (ESI MS) m/z: theor: 378.1831 found: 378.1834 ([M]$^{+\cdot}$ detected)

2.3.4. 2-(4-(Dimethylamino)Benzylidene)-1*H*-Indene-1,3(2*H*)-Dione **PP14**

4-Dimethylaminobenzaldehyde (0.38 g, 2.55 mmol), m_{exp} = 580 mg, 2.09 mmol, 82% yield). ^{1}H nmR (CDCl$_3$) δ: 3.14 (s, 6*H*), 6.75 (d, 2*H*, J = 9.2 Hz), 7.70–7.75 (m, 2*H*), 7.78 (s, 1*H*), 7.90–7.94 (m, 2*H*), 8.54 (d, 2*H*, J = 9.2 Hz); ^{13}C nmR (CDCl$_3$) δ: 40.1, 111.4, 122.1, 122.48, 122.49, 123.1, 134.1, 134.3, 137.9, 139.9, 142.3, 147.5, 154.0, 190.0, and 191.7; HRMS (ESI MS) m/z: theor: 277.1103 found: 277.1105 ([M]$^{+\cdot}$ detected).

2.3.5. 2-(2-Butoxy-4-(Diethylamino)Benzylidene)-1*H*-Indene-1,3(2*H*)-Dione **PP15**

2-Butoxy-4-(diethylamino)-benzaldehyde (0.64 g, 2.55 mmol), m_{exp} = 886 mg, 2.35 mmol, 92% yield. ^{1}H nmR (CDCl$_3$) δ: 1.03 (t, 3*H*, J = 7.4 Hz), 1.26 (t, 6*H*, J = 7.1 Hz), 1.60 (qt, 2*H*, J = 7.4 Hz), 1.90 (qt, 2*H*, J = 8.0 Hz), 3.48 (q, 4*H*, J = 7.1 Hz), 4.06 (t, 2*H*, J = 6.4 Hz), 6.03 (d, 1*H*, J = 1.7 Hz), 6.40 (dd, 1*H*, J = 9.3 Hz, J= 1.7 Hz), 7.66–7.70 (m, 2H), 7.85–7.89 (m, 2H), 8.48 (s, 1H), 9.35 (d, 1*H*, J = 9.3 Hz); ^{13}C nmR (CDCl$_3$) δ: 12.8, 13.9, 19.4, 31.1, 45.0, 68.2, 93.2, 104.9, 112.5, 121.1, 122.1, 122.2, 133.6, 133.9, 137.3, 139.9, 141.1, 142.2, 154.7, 163.7, 190.3, and 192.2; HRMS (ESI MS) m/z: theor: 377.1991 found: 377.1992 ([M]$^{+\cdot}$ detected).

2.3.6. 2-(4-(Diphenylamino)Benzylidene)-1*H*-Indene-1,3(2*H*)-Dione **PP16**

4-(Diphenylamino)benzaldehyde (0.70 g, 2.55 mmol) m_{exp} = 890 mg, 2.22 mmol, 87% yield. ^{1}H nmR (CDCl$_3$) δ: 7.01 (d, 2*H*, J = 8.9 Hz), 7.18-7.22 (m, 6H), 7.34–7.38 (m, 4H), 7.74–7.76 (m, 2H), 7.78 (s, 1H), 7.94–7.96 (m, 2H), 8.41 (d, 2*H*, J = 8.9 Hz); ^{13}C nmR (CDCl$_3$) δ: 118.9, 122.81, 122.84, 125.4, 125.5, 125.8, 126.6, 129.7, 134.5, 134.7, 136.8, 140.0, 142.4, 145.8, 146.5, 152.7, 189.6, and 191.2; HRMS (ESI MS) m/z: theor: 401.1416 found: 401.1418 ([M]$^{+\cdot}$ detected).

2.3.7. 2-(4-(Bis(4-Bromophenyl)Amino)Benzylidene)-1*H*-Indene-1,3(2*H*)-Dione **PP17**

4-(bis(4-bromophenyl)amino)benzaldehyde (1.1 g, 2.55 mmol), m_{exp} = 1.15 g, 2.06 mmol, 81% yield. ^{1}H nmR (CDCl$_3$) δ: 7.02–7.07 (m, 6*H*), 7.46 (d, 4*H*, J = 8.8 Hz), 7.76–7.78 (m, 3H), 7.96–7.98 (m, 2*H*), 8.42 (d, 2*H*, J = 8.9 Hz); ^{13}C nmR (CDCl$_3$) δ: 118.4, 120.0, 122.98, 123.01, 126.4, 126.9, 127.6, 133.0, 134.7, 135.0, 136.6, 140.1, 142.4, 144.8, 146.0, 151.5, 189.5, and 190.9; HRMS (ESI MS) m/z: theor: 556.9626 found: 556.9628 ([M]$^{+\cdot}$ detected).

2.3.8. 2-((9-Methyl-9H-Carbazol-3-Yl)Methylene)-1*H*-Indene-1,3(2*H*)-Dione **PP18**

9-Methyl-9*H*-carbazole-3-carbaldehyde (0.54 g, 2.55 mmol), m_{exp} = 671 mg, 1.99 mmol, 78% yield; ^{1}H nmR (CDCl$_3$) δ: 3.91 (s, 3*H*), 7.35 (t, 1*H*, J = 7.5 Hz), 7.43–7.48 (m, 2*H*), 7.54 (t, 1*H*, J = 7.3 Hz), 7.76–7.81 (m, 2*H*), 7.97–8.03 (m, 2*H*), 8.25 (d, 2*H*, J = 7.7 Hz), 8.69 (d, 1*H*, J = 8.7 Hz); ^{13}C nmR (CDCl$_3$) δ: 29.4, 108.7, 109.1, 120.7, 121.0, 122.9, 123.2, 123.6, 125.0, 125.6, 126.8, 128.6, 133.4, 134.6, 134.8, 140.0, 141.7, 142.5, 144.2, 148.8, 189.8, and 191.2; HRMS (ESI MS) m/z: theor: 337.1103 found: 337.1101 ([M]$^{+\cdot}$ detected).

2.3.9. 2-(3-(4-(Dimethylamino)Phenyl)Allylidene)-1*H*-Indene-1,3(2*H*)-Dione **PP19**

3-(4-(Dimethylamino)-phenyl)acrylaldehyde (0.45 g, 2.55 mmol), m_{exp} = 688 mg, 2.27 mmol, 89% yield. ^{1}H nmR (CDCl$_3$) δ: 3.09 (s, 6*H*), 6.69 (d, 2*H*, J = 8.9 Hz), 7.31 (d, 1*H*, J = 15.1 Hz), 7.59 (d, 2*H*, J = 8.4 Hz), 7.64 (d, 1*H*, J = 12.3 Hz), 7.71–7.74 (m, 2H), 7.90–7.92 (m, 2H), 8.28 (dd, 1*H*, J = 15.1 Hz, J = 12.3 Hz); ^{13}C nmR (CDCl$_3$) δ: 40.1, 111.9, 119.4, 122.4, 122.6, 123.8, 124.5, 131.3, 134.3, 134.4, 140.8, 142.1, 146.5, 152.5, 153.6, 190.8, and 191.2; HRMS (ESI MS) m/z: theor: 303.1259 found: 303.1255 ([M]$^{+\cdot}$ detected).

2.3.10. 2-(3,3-*Bis*(4-(Dimethylamino)Phenyl)Allylidene)-1*H*-Indene-1,3(2*H*)-Dione **PP20**

3-(4-(Dimethylamino)phenyl)-acrylaldehyde (0.75 g, 2.55 mmol), m_{exp} = 915 mg, 2.16 mmol, 85% yield. ^{1}H nmR (CDCl$_3$) δ: 3.06 (s, 6*H*), 3.07 (s, 6*H*), 6.67 (d, 2*H*, J = 9.0 Hz), 6.76 (d, 2*H*, J = 8.8 Hz), 7.20 (d, 2*H*, J = 8.7 Hz), 7.45 (d, 2*H*, J = 9.0 Hz), 7.67–7.71 (m, 3H), 7.84–7.89 (m, 2H), 8.28 (d, 1*H*, J = 12.8 Hz); ^{13}C nmR (CDCl$_3$) δ: 40.1, 40.2, 111.4, 111.5, 118.8, 122.1, 122.3, 123.9, 125.7, 128.8, 130.7, 132.1, 132.7, 133.4, 133.9, 134.1, 140.7, 142.0, 145.5, 151.7, 152.1, 165.5, 191.3, and 191.5; HRMS (ESI MS) m/z: theor: 422.1994 found: 422.1998 ([M]$^{+\cdot}$ detected).

3. Results and Discussion

3.1. Synthesis of the Dyes and Electron Acceptors

All dyes **PP1–PP20** presented in this work have been synthesized by a Knoevenagel reaction involving the aldehydes **D1–D10** and the two electron acceptors **EA4** and **EA1**, respectively (See Figure 3). The different reactions were performed in ethanol using piperidine as the catalyst. **PP1–PP20** were obtained with reaction yields ranging from 74% yield for **PP4** and **PP11** to 94% for **PP5** (see Table 1). All compounds were obtained as solids and they were characterized by ^{1}H, ^{13}C nmR spectroscopies, and HRMS spectrometry (see Supplementary Materials). It has to be noticed that the synthetic procedure to **EA4** has been greatly improved compared to that reported in the literature [47], enabling to reach a reaction yield of 91%. Indeed, a common method to synthesize 1,3-indanedione derivatives consists in a Claisen condensation of the corresponding diesters with ethyl acetate under basic conditions (generally sodium hydride).

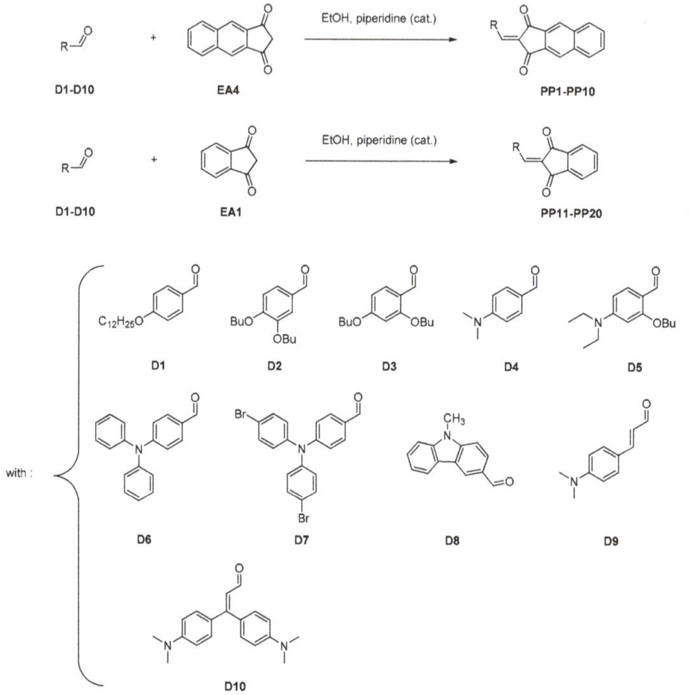

Figure 3. Synthetic pathways to **PP1–PP20** and the ten aldehydes used in this study.

Table 1. Reaction yields obtained for the synthesis of **PP1–PP20**.

Compounds	PP1	PP2	PP3	PP4	PP5	PP6	PP7	PP8	PP9	PP10
Reaction yields (%)	88	84	88	74	94	89	92	85	84	88
Compounds	PP11	PP12	PP13	PP14	PP15	PP16	PP17	PP18	PP19	PP20
Reaction yields (%)	74	85	75	82	92	87	81	78	89	85

The final product is obtained after the condensation step by decarboxylation of the intermediate salt under hot acidic conditions (see Figure 4). To access the starting compound, i.e., diethyl naphthalene-2,3-dicarboxylate, two different reaction pathways were examined, by esterification of naphthalene-2,3-dicarboxylic acid in ethanol in the presence of an excess of thionyl chloride, or by the classical esterification conditions consisting in refluxing the acid in ethanol in the presence of a catalytic amount of H_2SO_4 (see Figure 4). If diethyl naphthalene-2,3-dicarboxylate could be obtained in almost quantitative yields with the two procedures, our attempt to convert the diester obtained by the first procedure were unfruitful to form **EA4**. This is attributable to remaining traces of $SOCl_2$ in the diester, despite the numerous washings in basic conditions. Due to the presence of water traces in ethyl acetate, $SOCl_2$ could react with water to form HCl, neutralizing part of the NaH introduced. Conversely, in the second procedure, H_2SO_4 can be easily removed from the diester due to its catalytic use, avoiding this drawback. Using the oil obtained by this second procedure, a major improvement was obtained by replacing NaH 60% dispersion in oil by NaH 95% dispersion in oil for the first reaction step.

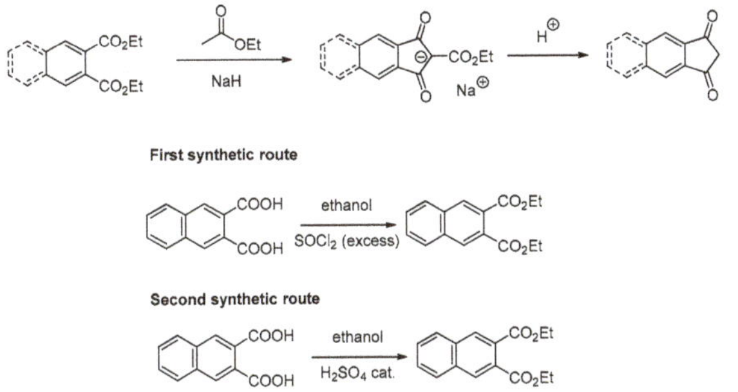

Figure 4. Synthetic route to 1,3-indanedione derivatives by the Claisen condensation and the two possible esterification procedures for naphthalene-2,3-dicarboxylic acid.

By using this more concentrated dispersion, the reaction yield of the crude materials for the two steps (the intermediate salt was not isolated) could be greatly increased. By modifying the decarboxylation time (increased from 1.5 h instead of 20 min), the quantity of NaH (2.5 eq. instead of 1.45 eq.) and the recrystallization solvent [47] (benzene replaced by the less toxic toluene) compared to that reported in the literature, the overall reaction yield after purification could be increased from 65% (literature) up to 91% in our optimized conditions.

3.2. Theoretical Calculations

Theoretical studies were realized in order to investigate the energy levels as well as the molecular orbitals (M.O.) compositions of the different dyes. DFT calculations of all synthetized compounds were performed by the B3LYP/6-311G(d,p) level of theory using Gaussian 09 programs. Dichloromethane as the solvent and the polarizable continuum model (PCM) as the solvent model were used for the TD-DFT calculations. The optimized geometries as well as the highest occupied molecular orbitals (HOMO) and the lowest unoccupied molecular orbital (LUMO), i.e., the frontier orbitals' electronic distributions of selected compounds (**PP6** and **PP16**) are given in Figure 5. The data for all compounds are supplied in the Supplementary Materials. The simulated absorption spectra are shown in the Figure 6 while the Table 2 summarizes all HOMO and LUMO energy levels and electronic transitions associated with the intramolecular charge transfer (ITC) absorption peaks. The assignment of the electronic transitions for λ_{max} reported in the Table 2 has been determined with GaussSum 3.0 software, and especially, the contribution of the different transitions.

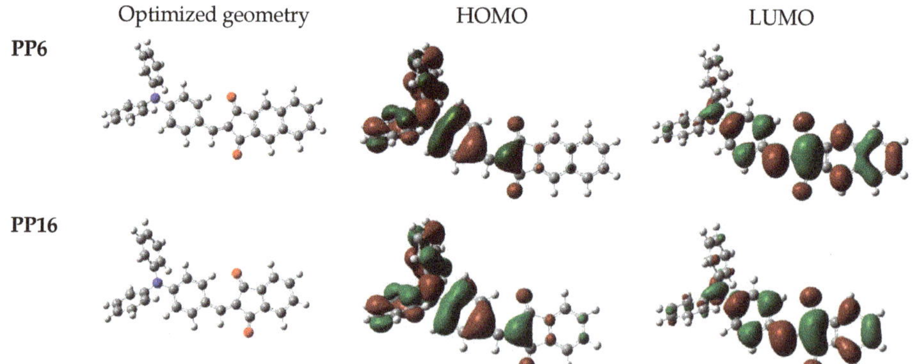

Figure 5. Optimized geometries and HOMO/LUMO electronic distributions of **PP6** and **PP16**.

As can be observed for all compounds, clear common trends are well observed for the two series:

- ICT Absorption:

The absorption peaks corresponded to the ICT are well observed in the Figure 6 and the values of maximal ICT absorption wavelengths are given in the Table 2. A clear trend is observed in both series: The ICT transition is mainly associated with the HOMO-LUMO transition. When the electron donor properties of the donor part increase, a clear redshift is obtained for the ICT transition. When we compare the two molecules in the two series with the same donor part, we observe that the ICT of the **EA4**-based molecule is redshift compared to that of **EA1**-based counterpart (for instance: λ_{max} (nm) = 518 and 490 for **PP6** and **PP16**, respectively, both compounds have triphenylamine as donor part).

- HOMO and LUMO Energy Levels:

Within the same series, the values of E_{HOMO} importantly vary as a function of the electron donating capacity of the donor part while the values of E_{LUMO} slightly change. When we compare cross the two series: two molecules with the same donor part, we observe that they have nearly the same E_{HOMO} while the E_{LUMO} of the **EA4**-based molecule is deeper compared to that of **EA1**-based counterpart, which is in good accordance with the more electron deficient nature of **EA4** (for instance: E_{HOMO} (eV) = −5.633 and −5.635 for **PP6** and **PP16**, respectively, while E_{LUMO} (eV) = −2.836 and −2.694, respectively. Both compounds have triphenylamine as donor part).

- HOMO and LUMO Orbitals Distribution:

The HOMO orbitals are well developed over the electron donor part while the LUMO orbitals are on the electron deficient moiety. Only a small overlap between the HOMO and LUMO orbitals is detected (see Figure 5).

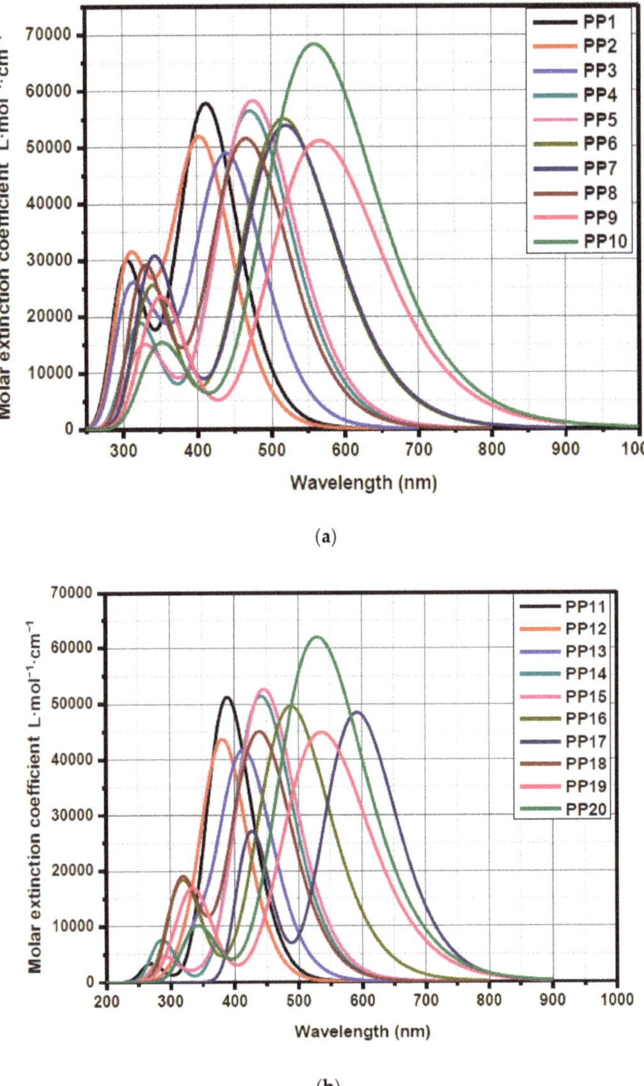

Figure 6. Simulated absorption spectra in dilute dichloromethane (5×10^{-3} M) of synthetized compounds **PP1–PP10** (**a**) and **PP11–PP20** (**b**).

Table 2. Summary of simulated absorption characteristics in dilute dichloromethane of synthetized compounds. Data were obtained in dichloromethane solution.

Compounds	E_{HOMO} (eV)	E_{LUMO} (eV)	λ_{max} (nm)	Transitions
PP1	−6.279	−2.865	414	HOMO->LUMO (95%)
PP2	−6.468	−2.983	406	HOMO->LUMO (72%) HOMO-1->LUMO (22%) HOMO-2->LUMO (5%)
PP3	−5.974	−2.691	441	HOMO->LUMO (98%)
PP4	−5.717	−2.668	473	HOMO->LUMO (99%)

Table 2. *Cont.*

Compounds	E_{HOMO} (eV)	E_{LUMO} (eV)	λ_{max} (nm)	Transitions
PP5	−5.522	−2.489	477	HOMO->LUMO (99%)
PP6	−5.633	−2.836	518	HOMO->LUMO (99%)
PP7	−5.826	−3.003	520	HOMO->LUMO (99%)
PP8	−5.953	−2.816	471	HOMO->LUMO (99%)
PP9	−5.554	−2.874	567	HOMO->LUMO (100%)
PP10	−5.262	−2.62	574	HOMO->LUMO (90%) HOMO-1->LUMO (10%)
PP11	−6.3	−2.71	390	HOMO->LUMO (97%)
PP12	−6.498	−2.842	385	HOMO->LUMO (94%) HOMO-1->LUMO (4%)
PP13	−5.982	−2.526	417	HOMO->LUMO (97%)
PP14	−5.715	−2.498	444	HOMO->LUMO (95%) HOMO->L+1 (4%)
PP15	−5.513	−2.309	448	HOMO->LUMO (95%) HOMO->LUMO+1 (4%)
PP16	−5.635	−2.694	490	HOMO->LUMO (99%)
PP17	−5.835	−2.878	493	HOMO->LUMO (99%)
PP18	−5.962	−2.656	443	HOMO->LUMO (98%)
PP19	−5.555	−2.748	537	HOMO->LUMO (100%)
PP20	−5.253	−2.489	545	HOMO->LUMO (85%) HOMO-1->LUMO (15%)

3.3. Optical Properties

All compounds were characterized by UV-visible spectroscopy and their absorption spectra in dichloromethane are provided in the Figure 7. **PP1–PP20** are push-pull compounds that possess an electron donating group and an electron accepting group which interact by mean of a π-conjugated system. The position of the charge transfer band will depend of parameters such as the strength of the electron donating/accepting groups, but also of the length of the π-conjugated system. By combining both effects, a decrease in the energy difference between the HOMO and the LUMO can be obtained, resulting in a shift of the absorption spectrum towards longer wavelengths.

While examining the first series **PP1–PP10**, absorption maxima ranging from 412 nm for **PP1** to 524 nm for **PP5** were found in dichloromethane for dyes exhibiting a short spacer. Elongation of the π-conjugated system resulted in a significant red-shift of the absorption maximum, shifting from 471 nm for **PP1** to 571 nm for **PP9**. The most red-shifted absorption was found for **PP10**, peaking at 603 nm. While comparing the **PP1–PP10** series with **PP11–PP20** based on 1,3-indanedione, absorption spectra of **PP11–PP20** were found to follow the same order than that of the **PP1–PP10** series, but with an absorption blue-shifted by about 30 nm (see Figure 7). Examination of the molar extinction coefficients for the two series also revealed the **PP1–PP10** series to exhibit higher molar extinction coefficients than that of the **PP11–PP20** series, consistent with an improvement of the molar absorptivity with the oscillator strength and the conjugation extension (see Figure 8) [48].

The experimental absorption spectra recorded in dichloromethane of all compounds are in good accordance with predicted properties obtained by DFT calculation. Notably, a good accordance between the theoretical absorption maxima can be found for all dyes (See Tables 2–4). Second, considering that the main absorption band was theoretically determined for all dyes originating from a HOMO->LUMO transition, this latter can thus be confidently assigned to the intramolecular charge transfer bands for all dyes. A contribution of the HOMO->LUMO transition to the ICT band ranging between 85% (for **PP20**) to 100% for **PP9** and **PP19** could be calculated.

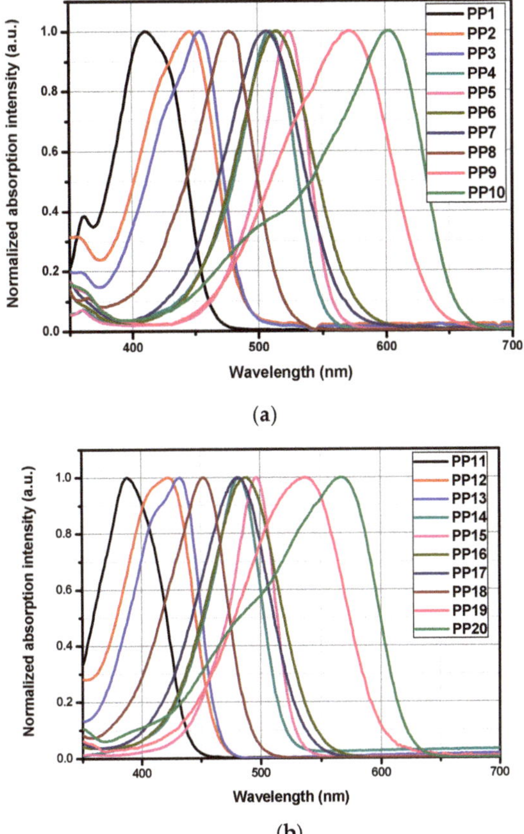

(**a**)

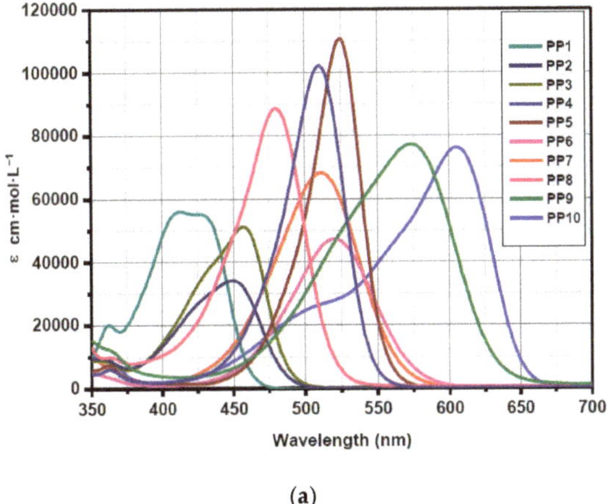

(**b**)

Figure 7. UV-visible absorption spectra of **PP1**–**PP10** (**a**) and **PP11**–**PP20** (**b**) in dichloromethane.

(**a**)

Figure 8. *Cont.*

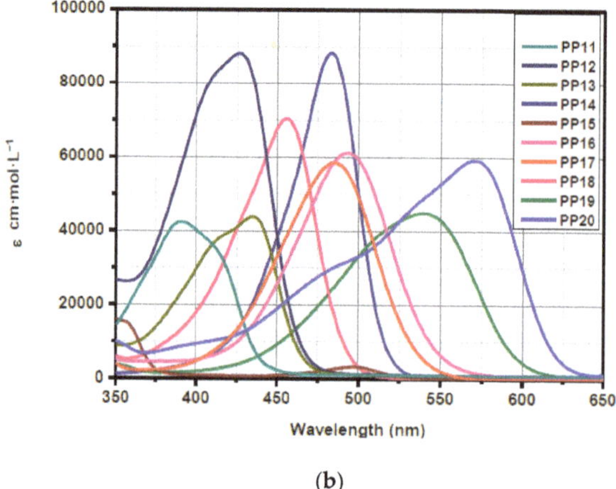

(b)

Figure 8. UV-visible absorption spectra of **PP1–PP10** (a) and **PP11–PP20** (b) in chloroform.

Table 3. Summary of the optical properties of **PP1–PP10** in twenty-three solvents and Kamlet and Taft parameters π^*.

Compounds	π^{*1}	PP1[2]	PP2[2]	PP3[2]	PP4[2]	PP5[2]	PP6[2]	PP7[2]	PP8[2]	PP9[2]	PP10[2]
diethyl ether	0.27	403	442	445	490	505	499	489	464	534	569
toluene	0.54	405	450	441	501	512	509	501	473	553	582
chloroform	0.78	412	450	457	520	524	520	510	480	574	606
THF	0.58	405	444	450	502	517	504	497	471	555	587
1,4-dioxane	0.55	402	443	446	498	512	503	496	468	545	581
acetone	0.71	405	441	449	505	520	501	496	471	562	595
dichloromethane	0.82	411	448	456	510	524	514	505	476	571	603
DMSO	1.00	414	447	458	522	534	512	506	482	590	620
heptane	-0.08	410	445	443	479	497	497	486	461	529	551
acetonitrile	0.75	404	439	450	507	523	505	496	471	565	597
dimethylformamide	0.87	409	447	454	514	528	509	503	478	579	609
ethyl acetate	0.54	404	444	447	498	513	500	491	468	546	579
p-xylene	0.43	406	448	448	496	509	504	497	470	547	580
1,2-dichloroethane	0.81	410	447	454	509	524	515	504	474	567	603
dimethylacetamide	0.88	410	447	456	516	529	508	502	477	577	608
diglyme	0.64	406	445	450	506	518	504	497	474	567	593
cyclohexane	0.00	412	446	445	482	499	502	492	462	532	556
triethylamine	0.14	439	454	450	486	501	503	507	463	533	559
hexane	-0.08	409	444	442	477	495	497	488	457	525	551
anisole	0.73	413	450	454	508	520	514	504	476	567	599
pentane	-0.09	408	442	440	476	495	495	485	458	524	548
nitrobenzene	1.01	433	452	461	518	532	519	509	484	585	613
diethyl carbonate	0.45	402	444	447	495	510	500	493	466	542	575

[1] Kamlet and Taft parameters; [2] Position of the ICT bands are given in nm.

Table 4. Summary of the optical properties of **PP11–PP20** in twenty-three solvents and Kamlet and Taft parameters π^*.

Compounds	π^{*1}	PP11[3]	PP12[3]	PP13[3]	PP14[3]	PP15[3]	PP16[3]	PP17[3]	PP18[3]	PP19[3]	PP20[3]
diethyl ether	0.27	384	404	424	462	480	474	467	442	502	531
toluene	0.54	388	408	429	474	486	485	476	451	520	552
chloroform	0.78	391	416	436	483	497	493	485	456	539	571
THF	0.58	387	411	428	476	490	479	472	448	518	553

Table 4. *Cont.*

Compounds	π^{*1}	PP11[3]	PP12[3]	PP13[3]	PP14[3]	PP15[3]	PP16[3]	PP17[3]	PP18[3]	PP19[3]	PP20[3]
1,4-dioxane	0.55	385	406	426	471	486	474	471	444	512	549
acetone	0.71	385	410	429	479	493	477	471	447	526	555
dichloromethane	0.82	388	417	433	482	497	489	482	452	537	567
DMSO	1.00	391	425	435	492	506	484	481	458	554	584
heptane	−0.08	379	400	422	457	473	476	468	439	478	523
acetonitrile	0.75	383	410	428	480	494	477	471	446	528	556
DMF	0.87	388	415	433	488	501	482	479	453	541	579
ethyl acetate	0.54	385	409	426	474	486	474	470	447	520	544
p-xylene	0.43	388	407	429	471	484	481	474	449	522	548
1,2-dichloroethane	0.81	388	421	432	481	496	487	480	453	536	567
dimethylacetamide	0.88	388	412	432	489	500	485	478	454	543	573
diglyme	0.64	386	412	429	479	490	480	472	450	524	559
cyclohexane	0.00	382	402	423	459	474	478	471	441	481	527
triethylamine	0.14	384	403	424	461	476	478	471	443	489	530
hexane	−0.08	378	400	422	455	471	474	467	437	476	521
anisole	0.73	389	415	432	481	495	487	482	454	530	570
pentane	−0.09	378	399	420	454	470	474	465	436	474	520
nitrobenzene	1.01	n.d.[2]	n.d. [2]	437	491	504	493	482	459	552	582
diethyl carbonate	0.45	384	405	426	471	484	475	470	445	510	541

[1] Kamlet and Taft parameters; [2] not determined. [3] Position of the ICT bands are given in nm.

3.4. Solvatochromism

All dyes **PP1–PP20** exhibited a good solubility in most of the common organic solvents so that examination of the solvatochromism could be carried out in 23 solvents of different polarities. It has to be noticed that alcohols such as methanol, ethanol, propan-2-ol, butan-1-ol and pentan-1-ol were initially considered as solvents for the solvatochromic study, but the absorption maxima obtained with these solvents were irregular compared to that obtained with the 23 other solvents. This specific behavior can be assigned to the fact that all dyes precipitated in alcohols such as in ethanol which was the solvent of reaction. Even if the absorption spectra could be recorded in alcohols for all dyes, presence of free molecules and aggregates in solution certainly modify the position of the absorption maxima. A summary of the absorption maxima for the twenty dyes are provided in Tables 3 and 4.

As evidenced in the Tables 3 and 4, analysis of the solvatochromism in solvents of different polarities confirmed the presence of an intramolecular charge transfer in all dyes. Intramolecular nature of the charge transfer was demonstrated by performing successive dilutions, intensity of the charge transfer band linearly decreasing with the dye concentrations. Various empirical polarity scales have been developed over the years to interpret the solvent-solute interaction and the Kamlet-Taft's [49], Dimroth-Reichardt's [50], Lippert-Mataga's [51], Catalan's [52], Kawski-Chamma-Viallet's [53], McRae's [54], Suppan's [55], and Bakhshiev's [56] scales can be cited as the most popular ones. Among all scales, the Kamlet-Taft solvent polarity scale proved to be the most adapted one, linear correlations being obtained for all dyes by plotting the absorption maximum vs. The empirical Taft parameters (see linear regression in Supplementary Materials). For all the other polarity scales based on the dielectric constant or the refractive index of solvents, no reasonable correlations could be established. The Kamlet and Taft equation is also a multiparametric equation that can take into account the dipolarity-polarizability (π^*), the hydrogen-donating and accepting ability (α and β) of the solvents, modelizing more precisely the interactions between the solvent and the solute. However, multiple linear regression analyses carried out on the triparametric Kamlet-Taft equation using the three solvent descriptors (α, β, π^*) did not improve the correlation coefficients, as demonstrated in the Table in SI. It can therefore be concluded that the dipolarity-polarizability of the solvent is the primary cause influencing the position of the ICT band.

As evidenced in the Figure 9 and Figures in Supplementary Materials, **PP1–PP20** show negative slopes with good linear correlations, indicative of a positive solvatochromism. Excepted for **PP1**, **PP2**, **PP11**, and **PP12** that possess weak electron donors, all dyes displayed strong negative slopes, indicative of a significant charge redistribution upon excitation. The most important solvatochromism was found for **PP9**, **PP10**, **PP19**, and **PP20** that exhibit the longest conjugated spacers.

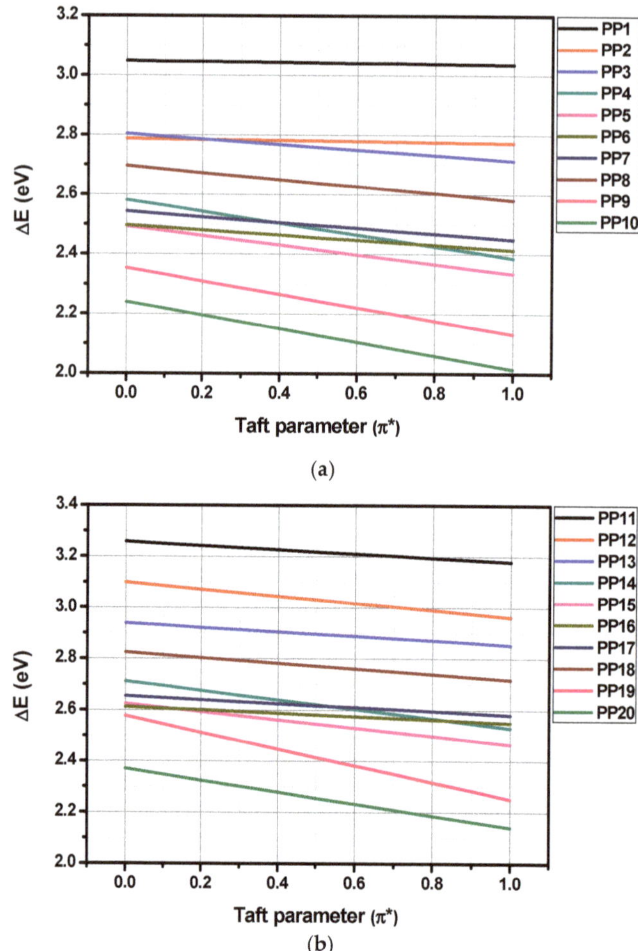

Figure 9. Variation of the positions of the charge transfer band with Kamlet-Taft empirical parameters for **PP1–PP10** (**a**) and **PP11–PP20** (**b**).

3.5. Fluorescence Spectroscopy

The fluorescent properties of all compounds were investigated in dichloromethane and toluene, in diluted solutions and the results are summarized in the Figure 10 and the Table 5. Interestingly, most of the chromophores were not emissive whatever the solvent is and this behavior is consistent with results classically reported in the literature for indane-1,3-dione derivatives [57–61]. However, it has to be noticed that a series of indane-1,3-dione derivatives was reported as being highly emissive, in a specific context, by use of oligo(phenylene)vinylene as electron donors [62]. Only eleven of these compounds displayed a weak emission in toluene so that the luminescence lifetime as well as the photoluminescence quantum yield were not determined for these compounds. While examining the emission maxima, the most red-shifted emissions were found for **PP9** and **PP10** displaying the most extended conjugated spacer but also the most extended electron acceptor. As attended, a blue-shifted emission was found for **PP20** (λ_{em} = 604 nm) relative to that of **PP10** (λ_{em} = 626 nm), directly resulting from its blue-shifted absorption maxima. While comparing the results obtained in dichloromethane and toluene, absorption maxima were found to be blue-shifted by *ca.* 20 nm for all dyes in toluene relative to that determined in dichloromethane. Based on the Taft parameters for both solvents (0.54 for toluene and 0.82 for dichloromethane), it can be concluded that a positive solvatochromism can also be observed in emission. This point is notably confirmed by the Stokes shifts determined in both dichloromethane and toluene, which are almost identical.

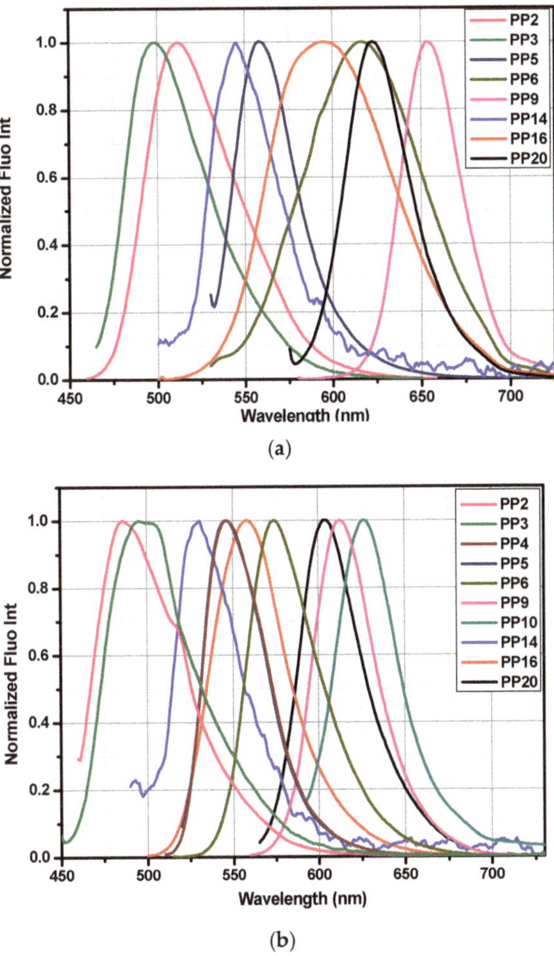

Figure 10. Fluorescence spectra of studied compounds in dichloromethane (**a**) and toluene (**b**).

Table 5. Fluorescence properties of the different compounds in dichloromethane and toluene solutions.

	Dichloromethane									
Compounds	PP1	PP2	PP3	PP4	PP5	PP6	PP7	PP8	PP9	PP10
excitation (nm)	411	448	456	510	524	524	505	476	571	603
emission (nm)	-	512	499	-	558	617	-	-	654	-
Stokes shift (nm)	-	64	43	-	34	103	-	-	-	-
Compounds	PP11	PP12	PP13	PP14	PP15	PP16	PP17	PP18	PP19	PP20
excitation (nm)	388	417	433	482	497	489	482	452	537	567
emission (nm)	-	-	-	545	-	596	-	-	-	623
Stokes shift (nm)	-	-	-	63	-	107	-	-	-	56
	Toluene									
Compounds	PP1	PP2	PP3	PP4	PP5	PP6	PP7	PP8	PP9	PP10
excitation (nm)	405	450	441	501	512	509	501	473	553	582
emission (nm)	-	487	495	546	546	574	-	-	613	626
Stokes shift (nm)	-	37	54	45	34	65	-	-	60	44
Compounds	PP11	PP12	PP13	PP14	PP15	PP16	PP17	PP18	PP19	PP20
excitation (nm)	388	408	429	474	486	485	476	451	520	552
emission (nm)	-	467	-	530	-	558	-	-	-	604
Stokes shift (nm)	-	59	-	56	-	73	-	-	-	52

3.6. Electrochemical Properties

The electrochemical properties of all compounds have been investigated by cyclic voltammetry (CV) in dilute solutions, either in acetonitrile or in dichloromethane. The selected voltammograms are shown in the Figure 11 and CV curves of all compounds are given in the Supplementary Materials. The redox potentials of all compounds are summarized in the Table 6 in which redox potentials are given against the half wave oxidation potential of the ferrocene/ferrocenium cation couple.

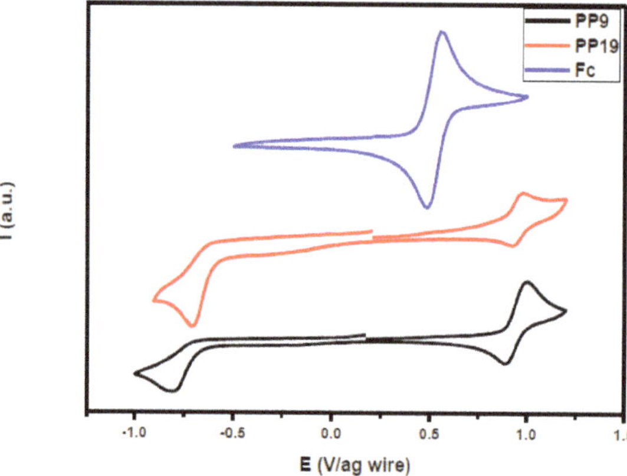

Figure 11. Selected examples of cyclic voltammograms of studied compounds (**PP9**, **PP19**) and Ferrocene (Fc).

Table 6. Electrochemical redox potentials of the studied compounds **PP1–PP20**.

Compounds	E_{red} (V/Fc)	E_{Ox} (V/Fc)	E_{HOMO} (eV)	E_{LUMO} (eV)	ΔE_{el} (eV)[1]	ΔE_{opt} (eV)[2]
PP1	−1.30	1.19	−5.99	−3.50	2.49	3.07
PP2	−1.26	1.18	−5.98	−3.54	2.44	2.82
PP3	−1.36	1.15	−5.95	−3.44	2.51	2.76

Table 6. *Cont.*

Compounds	E_{red} (V/Fc)	E_{Ox} (V/Fc)	E_{HOMO} (eV)	E_{LUMO} (eV)	ΔE_{el} (eV)[1]	ΔE_{opt} (eV)[2]
PP4	−1.33	0.68	−5.48	−3.47	2.01	2.45
PP5	−1.40	0.63	−5.43	−3.40	2.03	2.37
PP6	−1.32	0.79	−5.59	−3.48	2.11	2.46
PP7	−1.26	0.84	−5.64	−3.54	2.10	2.50
PP8	−1.38	0.95	−5.75	−3.42	2.33	2.63
PP9	−1.22	0.49	−5.29	−3.58	1.71	2.19
PP10	−1.30	0.41	−5.21	−3.50	1.71	2.08
PP11	−1.46	1.19	−5.99	−3.34	2.65	3.24
PP12	−1.44	1.18	−5.98	−3.36	2.62	3.02
PP13	−1.50	1.23	−6.03	−3.30	2.73	2.90
PP14	−1.56	0.65	−5.45	−3.24	2.21	2.58
PP15	−1.30	0.54	−5.34	−3.50	1.84	2.51
PP16	−1.44	0.77	−5.57	−3.36	2.21	2.60
PP17	−1.29	0.85	−5.65	−3.51	2.14	2.63
PP18	−1.52	0.94	−5.74	−3.28	2.46	2.78
PP19	−1.30	0.49	−5.29	−3.50	1.79	2.35
PP20	−1.30	0.36	−5.16	−3.50	1.66	2.23

[1] All potentials are recorded in 0.1 M $TBABF_4/CH_3CN$, except for **PP15** and **PP20** for which electrochemistry was carried out in 0.1 M $TBAClO_4/CH_2Cl_2$. E_{HOMO} (eV) = −4.8 − E_{ox} and E_{LUMO} (eV) = −4.8 − E_{red}; [2] Optical bandgaps determined in acetonitrile.

As shown in Figure 11, **PP9** and **PP19** differ from each other only in the nature of the accepting moiety. As expected, both compounds have quasi-reversible oxidation processes with identical oxidation potentials (Figure 11, Table 6). Indeed, the oxidation process for the two chromophores is centered onto the electron-donating part and this latter is the same for the two dyes. Conversely, the reduction potential of **PP9** comprising 1*H*-cyclopenta[*b*]naphthalene-1,3(2*H*)-dione (**EA4**) as the electron-accepting moiety is slightly lower than that of **PP19** (comprising **EA1** as the acceptor), leading to narrower electrochemical bandgap. This is in good accordance with the optical bandgap determined by UV-visible absorption spectroscopy, where **PP9** showed a red-shifted ICT band in comparison to **PP19** (See Figures 7 and 8).

The redox potentials of all other compounds are gathered in the Table 6. As can be seen in the same series (**PP1–PP10** and **PP11–PP20**) where the nature of the acceptor moiety is identical, their reduction potentials changed very slightly while the oxidation potential importantly vary as a function of the donor moiety.

The number and the substitution position of the alkoxy chains on the phenyl ring slightly influence the redox potentials of the targeted molecules (**PP1–3** and **PP11–13**). However, important variations were determined when the electron-donating ability of the electron donor was increased by the presence of diakylamino groups on the phenyl ring (see **PP4–5** and **PP14–15**). This phenomenon was even much more pronounced when a double bond was inserted between the donor and the acceptor moiety leading to more conjugated push pull molecules (see **PP9–10** & **PP19–20**). The presence of two 4-(*N,N*-methylamino)phenyl groups such as in **PP10** and **PP20** has only a negligible impact on the electrochemical property. In fact, while the second 4-(*N,N*-methylamino)phenyl group could increase the electron donating ability, examination of the mesomeric forms in **PP10** and **PP20** clearly evidences the two groups not to be able to contribute simultaneously to the electronic delocalization, as previously mentioned in the literature [46]. While comparing the dyes at identical electron donating groups, push pull compounds prepared with **EA4** (**PP1–PP10**) have lower reduction potentials than their counterparts **PP11–PP20** comprising **EA1** as the electron withdrawing groups. This is in perfect accordance with the higher electron accepting capacity of **EA4**.

The redox behaviors of synthesized molecules were then used to estimate their HOMO and LUMO energy levels by using the ferrocene (Fc) ionization potential value (4.8 eV vs. vacuum) as the standard. The correcting factor of 4.8 eV is based on calculations obtained by Pommerehne et al. [63]. It is also important to note that some other correcting factors have also been used in the literature [64]. The obtained values of E_{HOMO} and E_{LUMO} issued from electrochemical characterizations are summarized in the Table 6 for comparison, the optical bandgaps of all dyes in acetonitrile have been added. The Figure 12 shows a comparative presentation of the frontier orbitals' energy levels experimentally and theoretically obtained. We can see that the experimental findings fit well with the trend predicted by DFT calculation.

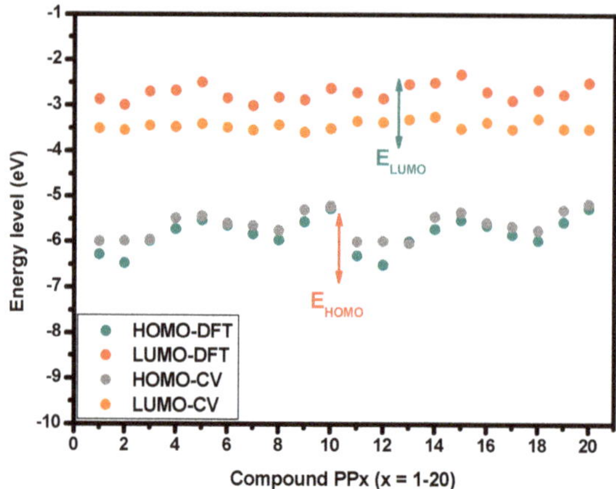

Figure 12. Comparison of frontier orbitals' energy levels obtained from cyclic voltammetry and DFT calculations.

4. Conclusions

In this work, a series of twenty dyes comprising indane-1,3-dione or 1*H*-cyclopenta[*b*] naphthalene-1,3(2*H*)-dione have been synthesized and examined for their photophysical properties. Introduction of a naphthalene moiety in **EA4** greatly contributed to improve the electron-withdrawing ability of the group. Notably, a red-shift of the intramolecular charge transfer band of ca. 30 nm was observed for all dyes prepared with this acceptor, compared to the parent series comprising **EA1**. A linear and positive solvatochromism was found for all dyes, demonstrating an important charge redistribution upon excitation. A good correlation between the experimental HOMO-LUMO gaps and the theoretical ones could be found. Examination of their electrochemical properties revealed these dyes to be reversibly oxidized whereas an irreversible reduction monoelectronic process was determined for all dyes. By extending the aromaticity of **EA4**, a significant red-shift of the absorption maximum could be obtained for all dyes compared their analogues comprising **EA1**. Future works will consist of further improving the electron accepting ability of **EA4**, which is achievable by functionalizing **EA4** with malononitrile. Based on the present work, the tetracyano derivatives of **EA4** will certainly allow the design of near-infrared dyes that can find applications in research fields such as telecommunications and defense.

Supplementary Materials: The following are available online at http://www.mdpi.com/1996-1944/12/8/1342/s1. [1]H and ^{13}C nmR spectra of all chromophores; Solvatochromism analyses in 23 solvents; contour plots of HOMO and LUMO energy levels of all dyes; cyclic voltammograms of all chromophores.

Author Contributions: C.P.; G.N. and F.D. directed the work and supervised the overall project. C.P., G.N. and F.D. carried out the synthesis of the chromophores, the chemical characterization of the dyes, the UV-visible absorption measurements and the solvatochromism experiments. T.-T.B. carried out the fluorescence and cyclic voltammetry experiments. S.P. carried out the theoretical calculations. Writing-Original Draft Preparation, C.P., G.N., F.D., T.-T.B. and S.P.; Writing-Review & Editing, C.P., G.N. and F.D.; M.N. and D.G. contributed to review the manuscript. All authors approved the final manuscript.

Materials **2019**, *12*, 1342

Funding: This research was funded by Aix Marseille University and The Centre National de la Recherche (CNRS) for financial supports. This research was also funded by the Agence Nationale de la Recherche (ANR agency) through the PhD grants of Corentin Pigot (ANR-17-CE08-0010 DUALITY project) and Guillaume Noirbent (ANR-17-CE08-0054 VISICAT project).

Conflicts of Interest: The authors declare no conflict of interest.

References

1. Raimundo, J.M.; Blanchard, P.; Gallego-Planas, N.; Mercier, N.; Ledoux-Rak, I.; Hierle, R.; Roncali, J. Design and synthesis of push-pull chromophores for second-order nonlinear optics derived from rigidified thiophene-based π-conjugating spacers. *J. Org. Chem.* **2002**, *67*, 205–218. [CrossRef]

2. El-Shishtawy, R.M.; Borbone, F.; Al-amshany, Z.M.; Tuzi, A.; Barsella, A.; Asiri, A.M.; Roviello, A. Thiazole azo dyes with lateral donor branch: Synthesis, structure and second order NLO properties. *Dyes Pig.* **2013**, *96*, 45–51. [CrossRef]

3. Paek, S.; Lee, J.K.; Ko, J. Synthesis and photovoltaic characteristics of push–pull organic semiconductors containing an electron-rich dithienosilole bridge for solution-processed small-molecule organic solar cells. *Sol. Energ. Mater. Solar Cells* **2014**, *120*, 209–217. [CrossRef]

4. Xu, S.J.; Zhou, Z.; Liu, W.; Zhang, Z.; Liu, F.; Yan, H.; Zhu, X. A twisted thieno[3,4-*b*]thiophene-based electron acceptor featuring a 14-π-electron indenoindene core for high-performance organic photovoltaics. *Adv. Mater.* **2017**, *29*, 1704510. [CrossRef]

5. Karak, S.; Liu, F.; Russell, T.P.; Duzhko, V.V. Bulk charge carrier transport in push-pull type organic semiconductor. *ACS Appl. Mater. Interfaces* **2014**, *6*, 20904–20912. [CrossRef] [PubMed]

6. Turkoglu, G.; Cinar, M.E.; Ozturk, T. Triarylborane-based materials for OLED applications. *Molecules* **2017**, *22*, 1522. [CrossRef] [PubMed]

7. Archetti, G.; Abbotto, A.; Wortmann, R. Effect of polarity and structural design on molecular photorefractive properties of heteroaromatic-based push-pull dyes. *Chem. Eur. J.* **2006**, *12*, 7151–7160. [CrossRef] [PubMed]

8. Ipuy, M.; Billon, C.; Micouin, G.; Samarut, J.; Andraud, C.; Bretonnière, Y. Fluorescent push–pull pH-responsive probes for ratiometric detection of intracellular pH. *Org. Biomol. Chem.* **2014**, *12*, 3641–3648. [CrossRef] [PubMed]

9. Gao, S.-H.; Xie, M.-S.; Wang, H.-X.; Niu, H.-Y.; Qu, G.-R.; Guo, H.-M. Highly selective detection of Hg^{2+} ion by push–pull-type purine nucleoside-based fluorescent sensor. *Tetrahedron* **2014**, *70*, 4929–4933. [CrossRef]

10. Jones, R.M.; Lu, L.; Helgeson, R.; Bergstedt, T.S.; McBranch, D.W.; Whitten, D.G. Building highly sensitive dye assemblies for biosensing from molecular building blocks. *Proc. Nat. Acad. Sci. USA* **2001**, *98*, 14769–14772. [CrossRef]

11. Ziessel, R.; Ulrich, G.; Harriman, A.; Alamiry, M.A.H.; Stewart, B.; Retailleau, P. Solid-state gas sensors developed from functional difluoroboradiazaindacene dyes. *Chem. Eur. J.* **2009**, *15*, 1359–1369. [CrossRef]

12. Tehfe, M.-A.; Dumur, F.; Graff, B.; Gigmes, D.; Fouassier, J.-P.; Lalevée, J. Blue-to-red light sensitive push-pull structured photoinitiators: indanedione derivatives for radical and cationic photopolymerization reactions. *Macromolecules* **2013**, *46*, 3332–3341. [CrossRef]

13. Tehfe, M.-A.; Dumur, F.; Graff, B.; Morlet-Savary, F.; Gigmes, D.; Fouassier, J.-P.; Lalevée, J. New push-pull dyes derived from Michler's ketone for polymerization reactions upon visible lights. *Macromolecules* **2013**, *46*, 3761–3770. [CrossRef]

14. Tehfe, M.-A.; Dumur, F.; Graff, B.; Morlet-Savary, F.; Gigmes, D.; Fouassier, J.-P.; Lalevée, J. Push-pull (thio)barbituric acid derivatives in dye photosensitized radical and cationic polymerization reactions under 457/473 nm Laser beams or blue LEDs. *Polym. Chem.* **2013**, *4*, 3866–3875. [CrossRef]

15. Mokbel, H.; Telitel, S.; Dumur, F.; Vidal, L.; Versace, D.-L.; Tehfe, M.-A.; Graff, B.; Toufaily, J.; Fouassier, J.-P.; Gigmes, D.; Hamieh, T.; Lalevée, J. Photoinitiating systems of polymerization and in-situ incorporation of metal nanoparticles in polymer matrixes upon visible lights: push-pull malonate and malonitrile based dyes. *Polym. Chem.* **2013**, *4*, 5679–5687. [CrossRef]

16. Xiao, P.; Frigoli, M.; Dumur, F.; Graff, B.; Gigmes, D.; Fouassier, J.-P.; Lalevée, J. Julolidine or fluorenone based push-pull dyes for polymerization upon soft polychromatic visible light or green light. *Macromolecules* **2014**, *47*, 106–112. [CrossRef]

17. Bures, F. Fundamental aspects of property tuning in push–pull molecules. *RSC Adv.* **2014**, *4*, 58826–58851. [CrossRef]

18. Noirbent, G.; Dumur, F. Recent advances on nitrofluorene derivatives: Versatile electron acceptors to create dyes absorbing from the visible to the near and far-infrared region. *Materials* **2018**, *11*, 2425. [CrossRef] [PubMed]

19. Solanke, P.; Ruzicka, A.; Mikysek, T.; Pytela, O.; Bures, F.; Klikar, M. From linear to T-shaped indan-1,3-dione push-pull molecules: A comparative study. *Helv. Chim. Acta* **2018**, *101*, e201800090. [CrossRef]

20. Solanke, P.; Bures, F.; Pytela, O.; Klikar, M.; Mikysek, T.; Mager, L.; Barsella, A.; Ruzicková, Z. T-shaped (donor-π-)2acceptor-π-donor push–pull systems based on indan-1,3-dione. *Eur. J. Org. Chem.* **2015**, 5339–5349. [CrossRef]

21. Swager, T.M.; Gutierrez, G.D. Push-pull chromophores from indan-1,3-dione. *Synfacts* **2014**, *10*, 0035.

22. Durand, R.J.; Gauthier, S.; Achelle, S.; Groizard, T.; Kahlal, S.; Saillard, J.-Y.; Barsella, A.; Le Poul, N.; Le Guen, F.R. Push–pull D–π-Ru–π-A chromophores: synthesis and electrochemical, photophysical and second order nonlinear optical properties. *Dalton Trans.* **2018**, *47*, 3965–3975. [CrossRef]

23. Nishiyabu, R.; Anzenbacher, P. 1,3-Indane-based chromogenic calixpyrroles with push-pull chromophores: Synthesis and anion sensing. *Org. Lett.* **2006**, *8*, 359–362. [CrossRef] [PubMed]

24. Cui, Y.; Ren, H.; Yu, J.; Wang, Z.; Qian, G. An indanone-based alkoxysilane dye with second order nonlinear optical properties. *Dyes Pigm.* **2009**, *81*, 53–57. [CrossRef]

25. Batsanov, A.S.; Bryce, M.R.; Davies, S.R.; Howard, J.A.K.; Whitehead, R.; Tanner, B.K. Studies on π-acceptor molecules containing the dicyanomethylene group. X-ray crystal structure of the charge-transfer complex of tetramethyltetrathiafulvalene and 2,3-dicyano-1,4-naphthoquinone: (TMTTF)$_3$-(DCNQ)$_2$. *J. Chem. Soc. Perkin Trans. 2* **1993**, *0*, 313–319. [CrossRef]

26. Wadsworth, A.; Moser, M.; Marks, A.; Little, M.S.; Gasparini, N.; Brabec, C.J.; Baran, D.; McCulloch, I. Critical review of the molecular design progress in non-fullerene electron acceptors towards commercially viable organic solar cells. *Chem. Soc. Rev.* **2019**. [CrossRef] [PubMed]

27. Liu, Y.; Li, M.; Zhou, X.; Jia, Q.-Q.; Feng, S.; Jiang, P.; Xu, X.; Ma, W.; Li, H.-B.; Bo, Z. Nonfullerene acceptors with enhanced solubility and ordered packing for high-efficiency polymer solar cells. *ACS Energy Lett.* **2018**, *3*, 1832–1839. [CrossRef]

28. Sanguinet, L.; Williams, J.C.; Yang, Z.; Twieg, R.J.; Mao, G.; Singer, K.D.; Wiggers, G.; Petschek, R.G. Synthesis and characterization of new truxenones for nonlinear optical applications. *Chem. Mater.* **2006**, *18*, 4259–4269. [CrossRef]

29. Sigalov, M.V.; Shainyan, B.A.; Chipanina, N.N.; Oznobikhina, L.P. Intra- and intermolecular hydrogen bonds in pyrrolylindandione derivatives and their interaction with fluoride and acetate: Possible anion sensing properties. *J. Phys. Chem. A* **2013**, *117*, 11346–11356. [CrossRef]

30. Feng, H.; Qiu, N.; Wang, X.; Wang, Y.; Kan, B.; Wan, X.; Zhang, M.; Xia, A.; Li, C.; Liu, F.; Zhang, H.; Chen, Y. An A-D-A type small-molecule electron acceptor with end-extended conjugation for high performance organic solar cells. *Chem. Mater.* **2017**, *29*, 7908–7917. [CrossRef]

31. Li, R.; Liu, G.; Xiao, M.; Yang, X.; Liu, X.; Wang, Z.; Ying, L.; Huang, F.; Cao, Y. Non-fullerene acceptors based on fused-ring oligomers for efficient polymer solar cells via complementary light-absorption. *J. Mater. Chem. A* **2017**, *5*, 23926–23936. [CrossRef]

32. Frisch, M.J.; Trucks, G.W.; Schlegel, H.B.; Scuseria, G.E.; Robb, M.A.; Cheeseman, J.R.; Montgomery, J.A., Jr.; Vreven, T.; Kudin, K.N.; Burant, J.C.; et al. *GAUSSIAN Program*; Gaussian, Inc.: Wallingford, CT, USA, 2009.

33. Lee, C.; Yang, W.; Parr, R.G. Development of the Colle-Salvetti correlation-energy formula into a functional of the electron density. *Phys. Rev. B. Condens. Matter.* **1988**, *37*, 785–789. [CrossRef]

34. Becke, A.D. A new mixing of Hartree–Fock and local density-functional theories. *J. Chem. Phys.* **1993**, *98*, 1372–1377. [CrossRef]

35. Hehre, W.J.; Ditchfield, R.; Pople, J.A. Self-consistent molecular orbital methods. XII. Further extensions of Gaussian-type basis sets for use in molecular orbital studies of organic molecules. *J. Chem. Phys.* **1972**, *56*, 2257–2261. [CrossRef]

36. Tomasi, J.; Mennucci, B.; Cances, E. The IEF version of the PCM solvation method: an overview of a new method addressed to study molecular solutes at the QM ab initio level. *J. Mol. Struct. THEOCHEM* **1999**, *464*, 211–226. [CrossRef]

37. Scalmani, G.; Frisch, M.J. Continuous surface charge polarizable continuum models of solvation. I. General formalism. *J. Chem. Phys.* **2010**, *132*, 114110. [CrossRef] [PubMed]

38. O'Boyle, N.M.; Tenderholt, A.L.; Langner, K.M. cclib: A library for package-independent computational chemistry algorithms. *J. Comp. Chem.* **2008**, *29*, 839–845. [CrossRef]

39. Delong, W.; Lanying, W.; Yongling, W.; Shuang, S.; Juntao, F.; Xing, Z. Natural α-methylenelactam analogues: Design, synthesis and evaluation of α-alkenyl-γ and δ-lactams as potential antifungal agents against Colletotrichum orbiculare. *Eur. J. Med. Chem.* **2017**, *130*, 286–307. [CrossRef]

40. Zhang, Q.; Peng, H.; Zhang, G.; Lu, Q.; Chang, J.; Dong, Y.; Shi, X.; Wei, J. facile bottom-up synthesis of coronene-based 3-fold symmetrical and highly substituted nanographenes from simple aromatics. *J. Am. Chem. Soc.* **2014**, *136*, 5057–5064. [CrossRef] [PubMed]

41. Pollard, A.J.; Perkins, E.W.; Smith, N.A.; Saywell, A.; Goretzki, G.; Phillips, A.G.; Argent, S.P.; Sachdev, H.; Mueller, F.; Huefner, S.; et al. Supramolecular Assemblies Formed on an Epitaxial Graphene Superstructure. *Angew. Chem. Int. Ed.* **2010**, *49*, 1794–1799. [CrossRef]

42. He, F.; Tian, L.; Tian, X.; Xu, H.; Wang, Y.; Xie, W.; Hanif, M.; Xia, J.; Shen, F.; Yang, B.; et al. diphenylamine-substituted cruciform oligo(phenylene vinylene): enhanced one- and two-photon excited fluorescence in the solid state. *Adv. Funct. Mater.* **2007**, *17*, 1551–1557. [CrossRef]

43. Wang, H.; Ding, W.; Wang, G.; Pan, C.; Duan, M.; Yu, G. Tunable molecular weights of poly(triphenylamine-2,2'-bithiophene) and their effects on photovoltaic performance as sensitizers for dye-sensitized solar cells. *J. Appl. Polym. Sci.* **2016**, *133*, 44182. [CrossRef]

44. Al Mousawi, A.; Dumur, F.; Garra, P.; Toufaily, J.; Hamieh, T.; Graff, B.; Gigmes, D.; Fouassier, J.-P.; Lalevée, J. carbazole scaffold based photoinitiator/photoredox catalysts: toward new high performance photoinitiating systems and application in led projector 3d printing resins. *Macromolecules* **2017**, *50*, 2747–2758. [CrossRef]

45. Malina, I.; Kampars, V.; Turovska, B.; Belyakov, S. Novel green-yellow-orange-red light emitting donor-π-acceptor type dyes based on 1,3-indandione and dimedone moieties. *Dyes Pigm.* **2017**, *139*, 820–830. [CrossRef]

46. Dumur, F.; Mayer, C.R.; Dumas, E.; Miomandre, F.; Frigoli, M.; Sécheresse, F. New chelating stilbazonium-like dyes from michler's ketone. *Org. Lett.* **2008**, *10*, 321–324. [CrossRef]

47. Buckle, D.R.; Morgan, N.J.; Ross, J.W.; Smith, H.; Spicer, B.A. Antiallergic activity of 2-nitroindan-1, 3-diones. *J. Med. Chem.* **1973**, *16*, 1334–1339. [CrossRef]

48. Lee, Y.; Jo, A.; Park, S.B. Rational Improvement of Molar Absorptivity Guided by Oscillator Strength: A Case Study with Furoindolizine-Based Core Skeleton. *Angew. Chem.* **2015**, *127*, 15915–15919. [CrossRef]

49. Kamlet, M.J.; Abboud, J.-L.M.; Abraham, M.H.; Taft, R.W. Linear solvation energy relationships. 23. A comprehensive collection of the solvatochromic parameters, .pi.*, .alpha., and .beta., and some methods for simplifying the generalized solvatochromic equation. *J. Org. Chem.* **1983**, *48*, 2877–2887. [CrossRef]

50. Reichardt, C. Solvatochromic Dyes as Solvent Polarity Indicators. *Chem. Rev.* **1994**, *94*, 2319–2358. [CrossRef]

51. Lippert, E.Z. Dipolmoment und Elektronenstruktur von angeregten Molekülen. *Naturforsch.* **1955**, *10a*, 541–545. [CrossRef]

52. Catalan, J. On the ET (30), π*, Py, S', and SPP empirical scales as descriptors of nonspecific solvent effects. *J. Org. Chem.* **1997**, *62*, 8231–8234. [CrossRef]

53. Kawski, A. Zur Lösungsmittelabhängigkeit der Wellenzahl von Elektronenbanden Lumineszierender Moleküle und über die Bestimmung der Elektrischen Dipolmomente im Anregungszustand. *Acta Phys. Polon.* **1966**, *29*, 507–518.

54. McRae, E.G. Theory of Solvent Effects on Molecular Electronic Spectra. *Frequency Shifts. J. Phys. Chem.* **1957**, *61*, 562–572. [CrossRef]

55. Suppan, P. Solvent effects on the energy of electronic transitions: experimental observations and applications to structural problems of excited molecules. *J. Chem. Soc. A* **1968**, *0*, 3125–3133. [CrossRef]

56. Bakshiev, N.G. Universal intermolecular interactions and their effect on the position of the electronic spectra of molecules in two component solutions. *Opt. Spektrosk.* **1964**, *16*, 821–832.

57. Redoa, S.; Eucat, G.; Ipuy, M.; Jeanneau, E.; Gautier-Luneau, I.; Ibanez, A.; Andraud, C.; Bretonnière, Y. Tuning the solid-state emission of small push-pull dipolar dyes to the far-red through variation of the electron-acceptor group. *Dyes Pigm.* **2018**, *156*, 116–132.

58. Jeong, H.; Chitumalla, R.K.; Kim, D.W.; Prabhakar Vattikuti, S.V.; Thogiti, S.; Cheruku, R.; Kim, J.H.; Jang, J.; Koyyada, G.; Jung, J.H. The comparative study of new carboxylated 1,3-indanedione sensitizers with standard cyanoacetic acid dyes using co-adsorbents in dye-sensitized solar cells. *Chem. Phys. Lett.* **2019**, *715*, 84–90. [CrossRef]

59. Liu, Y.; Sun, Y.; Li, M.; Feng, H.; Ni, W.; Zhang, H.; Wan, X.; Chen, Y. Efficient carbazole-based small-molecule organic solar cells with an improved fill factor. *RSC Adv.* **2018**, *8*, 4867–4871. [CrossRef]

60. Cao, D.; Peng, J.; Hong, Y.; Fang, X.; Wang, L.; Meier, H. Enhanced performance of the dye-sensitized solar cells with phenothiazine-based dyes containing double D–A branches. *Org. Lett.* **2011**, *13*, 1610–1613. [CrossRef] [PubMed]

61. Dai, S.; Zhao, F.; Zhang, Q.; Lau, T.-K.; Li, T.; Liu, K.; Ling, Q.; Wang, C.; Lu, X.; You, W.; Zhan, X. Fused nonacyclic electron acceptors for efficient polymer solar cells. *J. Am. Chem. Soc.* **2017**, *139*, 1336–1343. [CrossRef]

62. Guerlin, A.; Dumur, F.; Dumas, E.; Miomandre, F.; Wantz, G.; Mayer, C.R. Tunable optical properties of chromophores derived from oligo(para-phenylene vinylene). *Org. Lett.* **2010**, *12*, 2382–2385. [CrossRef]

63. Pommerehne, J.; Vestweber, H.; Guss, W.; Mahrt, R.F.; Bässler, H.; Porsch, M.; Daub, J. Efficient two layer leds on a polymer blend basis. *Adv. Mater.* **1995**, *7*, 551–554. [CrossRef]

64. Cardona, C.M.; Li, W.; Kaifer, A.E.; Stockdale, D.; Bazan, G.C. Electrochemical considerations for determining absolute frontier orbital energy levels of conjugated polymers for solar cell applications. *Adv. Mater.* **2011**, *23*, 2367–2371. [CrossRef] [PubMed]

 *materials*

Article

Optoelectronic Properties of C_{60} and C_{70} Fullerene Derivatives: Designing and Evaluating Novel Candidates for Efficient P3HT Polymer Solar Cells

Juganta K. Roy [†], Supratik Kar [*,†] and Jerzy Leszczynski [*]

Interdisciplinary Center for Nanotoxicity, Department of Chemistry, Physics and Atmospheric Sciences, Jackson State University, Jackson, MS 39217, USA

* Correspondence: supratik.kar@icnanotox.org (S.K.); jerzy@icnanotox.org (J.L.);
  Tel.: +1-601-979-0253 (S.K.); Fax: +1-601-979-7823 (J.L.)

† These authors contributed equally to this work.

Received: 24 June 2019; Accepted: 12 July 2019; Published: 16 July 2019

Abstract: Ten novel fullerene-derivatives (FDs) of C_{60} and C_{70} had been designed as acceptor for polymer solar cell (PSC) by employing the quantitative structure-property relationship (QSPR) model, which was developed strategically with a reasonably big pool of experimental power conversion efficiency (PCE) data. The QSPR model was checked and validated with stringent parameter and reliability of predicted PCE values of all designed FDs. They were assessed by the applicability domain (AD) and process randomization test. The predicted PCE of FDs range from 7.96 to 23.01. The obtained encouraging results led us to the additional theoretical analysis of the energetics and UV-Vis spectra of isolated dyes employing Density functional theory (DFT) and Time-dependent-DFT (TD-DFT) calculations using PBE/6-31G(d,p) and CAM-B3LYP/6-311G(d,p) level calculations, respectively. The FD4 is the best C_{60}-derivatives candidates for PSCs as it has the lowest exciton binding energy, up-shifted lowest unoccupied molecular orbital (LUMO) energy level to increase open-circuit voltage (V_{OC}) and strong absorption in the UV region. In case of C_{70}-derivatives, FD7 is potential candidate for future PSCs due to its strong absorption in UV-Vis region and lower exciton binding energy with higher V_{OC}. Our optoelectronic results strongly support the developed QSPR model equation. Analyzing QSPR model and optoelectronic parameters, we concluded that the FD1, FD2, FD4, and FD10 are the most potential candidates for acceptor fragment of fullerene-based PSC. The outcomes of tactical molecular design followed by the investigation of optoelectronic features are suggested to be employed as a significant resource for the synthesis of FDs as an acceptor of PSCs.

Keywords: DFT; fullerene derivative; P3HT; polymer solar cell; QSPR; TD-DFT

1. Introduction

Polymer solar cell (PSC) is a subject of discussion over the last decade due to its initial encouraging power conversion efficiency (PCE). Over time, organic dye-sensitized solar cells (DSSCs) and perovskite solar systems outperform the PSCs based on better and efficient PCE. Additionally, high cost and low-life time factors are other threats, which pose a great task for the researchers [1–4]. The PSC functions similarly to another kind of solar cell through the conversion of photons into an electrical current. The most common types of PSC are fullerene-based and non-fullerene based where they act as acceptor fragment and the role of polymer is a donor. Cuiet al. [2] experimentally showed that chlorinated non-fullerene acceptor-based PSC converts 16.5% of solar energy into an electrical current which is the highest reported PCE value for any non-fullerene PSCs till date. On the contrary, Meng et al. [4] reported 17.3% conversion employing tandem cell strategy using PTB7-Th:O6T-4F:PC_{71}BM (PTB7-Th is poly[4,8-bis(5-(2-ethylhexyl)thiophen-2-yl)benzo[1,2-b:4,5-b′]

dithiophene-co-3-fluorothieno[3,4-b]thiophene-2-carboxylate]; O6T-4F is carbon-oxygen-bridged i8 difluoro-substituted 1,1-dicyanomethylene-3-indanone; $PC_{71}BM$ is [6,6]-phenyl C71 butyric methyl ester) with an architecture of ITO/ZnO/PFN-Br/active-layer/M-PEDOT/Ag and ITO/ZnO/active-layer/MoO3/Ag where PC71BM acts as acceptor material under fullerene-based PSC. Fullerene derivative (FD) $PC_{61}BM$ conjugated with diverse polymers P3HT (poly-(3-hexyl)thiophene), PTPTB (poly-N-dodecyl-2,5,-bis(2'-thienyl)pyrrole, 2,1,3-benzothiadiazole), PEOPT (poly(3-(4'-(1'',4'',7''-trioxaoctyl) phenyl)thiophene), PFDTBT (poly{[2,7-(9-(20-ethylhexyl)-9-hexylfluorene])-alt-[5,50-(40,70-di-2-thienyl-20,10,30-benzothiadiazole)]}) showed PCE values range from 2.8 to 4.4 [5–8], 1.7–2.1 [9], 1.75 [9] 1.9 [9], respectively. Even, PCPDTBT (poly[2,6-(4,4-bis(2-ethylhexyl)-4H-cyclopenta[2,1-b;3,4-b']dithiophene)-alt-4,7(2,1,3-benzothiadiazole)]) and PBDTP-DTBT (poly{4,8-bis(4-(2-ethylhexyl)- phenyl)-benzo[1,2-b:4,5-b']dithiophene-alt-[4,7-di(4-(2-ethylhexyl)-2- thienyl)-2,1,3-benzothiadiazole)-5, 5'-diyl]}) showed PCE of 3.2 [10] and 8.07 [11] with PC71BM, respectively. While, PCDTBT (poly[N-9'-heptadecanyl-2,7-carbazole-alt-5,5-(4',7'-di-2-thienyl-2',1',3'-benzothiadiazole)]) and PCPDTBT offers PCE of 6.1 [12] and 6.16 [13] with $PC_{70}BM$, respectively.

The PCE value of existing PSCs is reasonable but not better than other commercial solar cell systems. Thus, improved and efficient PSCs are required by the implementation of rational designing of different fragments followed by optoelectronic properties evaluation to establish them as a future system. Based on the above discussion, it is quite clear that a good amount of simple and complex polymers has been examined, but the PCE value has not improved drastically. Therefore, we have aimed for novel modifications of FDs to improve the PCE of PSCs. Arbitrarily synthesizing various FDs is not a practical solution, as well as it is costly and time-consuming. Thus, considering experimental PCE data of existing FDs, in silico models can be prepared by quantitative structure-property relationship (QSPR) model [14]. Our group has proposed the first QSPR model followed by the virtual screening of FDs generating future lead acceptor fragment for PSC with PCE value of 12.11% [15]. Additionally, the QSPR technique was successfully employed in all steps from designing to prediction purpose for DSSCs by our group [16,17] and other researchers [18,19] with encouraging outcomes. Therefore, without any doubt, the QSPR modeling can be tactfully employed for designing better acceptor FDs for PSCs.

In continuation of our previous work [15], in the present study, we have prepared a QSPR model followed by implementation of the mechanistic interpretation and identified vital structural fragments obtained from the model to design new FDs as an acceptor for PSC. Previously we had used the QSPR model to screen 169 FDs to find the best FDs based on PCE value only without considering identified features from the model [15]. In the present study, designing will help us to consider the structural fragments more precisely and effectively. Ten FDs have been designed including seven C_{60} and three C_{70} FDs. The PCE of designed FDs are predicted employing the developed QSPR model prepared from 59 existing experimental PCE data of FDs. Top four lead acceptors (3 from C_{60} and 1 C_{70}) are further employed for energetics study along with analysis of UV-Vis spectra of isolated dyes. The rational scheme from designing to the electrochemical analysis of FDs for PSCs offers a detailed idea of how one can implement a QSPR model to design various future solar cells, not confined to only PSCs pool of species.

2. Materials and Methods

2.1. QSPR Modeling Study and Designing

2.1.1. Dataset

A series of 59 FDs consist of 52 C_{60} and 7 C_{70} derivatives as acceptor for PSCs with experimental PCE data were taken from our previous research [15] to generate a statistically acceptable and predictive QSPR model. The experimental data for all solar cells is measured according to bulk-heterojunction (BHJ) devices, where FD acts as the electron acceptor and Poly(3-hexylthiophene) (P3HT), a commonly used photovoltaic polymer as the donor material. The experimental data and chemical structures of FDs have been reported in Table S1 in the Supplementary material section.

2.1.2. Descriptors Calculation

Molecular structures of FDs had been drawn in GaussView 6.0 [20] and optimized by semi-empirical PM6 method using Gaussian 16 software [21]. The output structures were employed in MarvinView (ChemAxon) [22] software to calculate physicochemical and solvent accessible surface area descriptors. To compute Simplex Representation of Molecular Structure (SiRMS) [23,24] descriptors, QSAR4U software was applied which helps to identify and understand the major structural fragments responsible for higher PCE.

2.1.3. Dataset Division and Modeling Tools

The entire pool of descriptors was treated with a 0.0001 variance cutoff and passed through a 0.99 correlation coefficient to eradicate the highly correlated feature and decrease the noise level among descriptors. Followed by the dataset is divided into 3:1 ratio randomly by generating training and test sets with 44 and 15 FDs, respectively. The training set was then employed to develop a PLS based model using Partial Least Squares version 1.0 tool [25].

2.1.4. Model Validation and Designing Criteria

To assess the quality of a QSPR model followed by its prediction capability towards new compounds depend largely on statistical metrics. Internal metrics like R^2 (goodness-of-fit) and leave-one-out cross-validation (Q^2_{LOO}) are important parameters. While external validation metric R^2_{pred} or $Q^2_{ext(F1)}$ signify the predictability. Along with these classical parameters, to check the quality of the developed model, we have further employed stringent metrics like the r_m^2 metrics [26], the $Q^2_{ext(F2)}$ [27] and Golbraikh and Tropsha's [28] criteria. To follow the Organization for Economic Co-operation and Development (OECD) principle 3, we have studied the applicability domain test employing the Euclidean distance approach [29]. To check the robustness of the model, Y-randomization technique had been performed to generate 100 random models [30]. The average R^2 and Q^2_{LOO} values of all 100 random models should be failed the stipulated threshold value of 0.5.

2.2. Quantum Study of Designed FDs

To model P3HT polymer, an oligomer with 8 monomers considered to study the compatibility of the designed four FDs [31,32]. In all of the computations, PCBM was used as a reference. A variety of functionals [B3LYP, CAM-B3LYP, PBE] and basis sets [6-31G(d,p), 6-311G(d,p)] were used for the accurate description of the frontier orbitals of [6,6]-phenyl-C61-butyricacidmethylester (PCBM) and Poly(3-hexylthiophene) (P3HT) (see Table 1). PBE/6-31G(d,p) and CAM-B3LYP/6-311G(d,p) level of theory employed for the DFT and TDFT calculations, respectively. Absorption spectra of FDs in chlorobenzene solvent have been simulated by the Conductor like polarizable continuum model (C-PCM) [33] considering 20 low-lying singlet-singlet allowed transitions. All calculations were performed with the Gaussian 16 program package [21].

Table 1. Energy profiles highest occupied molecular orbital (HOMO), lowest unoccupied molecular orbital (LUMO) and their gap (E_{gap}) of the isolated P3HT oligomer, and PCBM. All the energies are in eV and PCBM/P3HT.

Method.	E_{HOMO}	E_{LUMO}	E_{GAP}
PBE/6-31G(d,p)	−5.94/−4.00	−3.12/−2.46	2.82/1.54
PBE/6-311G(d,p)	−6.24/−4.49	−3.45/−2.43	2.79/2.06
B3LYP/6-31G(d,p)	−5.63/−4.76	−3.06/−1.81	2.57/2.95
B3LYP/6-311G(d,p)	−6.02/−5.08	−3.47/−1.92	2.55/3.16
CAMB3LYP/6-31G(d,p)	−6.78/−6.16	−2.08/−0.53	4.70/5.63
CAMB3LYP/6-311G(d,p)	−7.17/−6.36	−2.51/−0.77	4.66/5.59
Experiment [34]	6.0/5.2	4.2/3.2	1.8/2.0

3. Results and Discussion

3.1. QSPR Modeling

The training set offered 7 features (descriptors) and 6 latent variables (LVs) based PLS equation (Equation (1)). To judge the goodness-of-quality of the presented equation along with predictivity of the test set molecules, we have checked a series of stringent statistical metrics and all of them passed the stipulated threshold values (Table 2).

$$
\begin{aligned}
\text{PCE (\%)} = &\ 2.50 + 1.84*S\_A(chg)/A\_D\_D\_D/1\_2s,1\_3s,3\_4a/6 \\
&-0.78*Fr5(chg)/B\_C\_C\_C\_D/1\_4s,2\_3s,2\_4s,3\_4s/ \\
&-0.06*Fr5(type)/C.3\_C.3\_C.3\_C.3\_H/1\_2s,2\_3s,3\_4s,4\_5s/ \\
&-0.937*Fr5(att)/C\_C\_E\_E\_E/1\_3s,2\_4s,3\_5a,4\_5a/ \\
&-0.11*Fr5(type)/C.3\_C.3\_C.AR\_C.AR\_C.AR/1\_4s,2\_3s,2\_5s,4\_5a/ \\
&+0.61*S\_A(type)/C.3\_C.3\_C.3\_C.AR/1\_3s,2\_3s,3\_4s/5-0.008*ASA\_P
\end{aligned}
\tag{1}
$$

Table 2. Obtained statistical data from the developed quantitative structure-property relationship (QSPR) model.

Validation	Metrics	Value	Threshold
	$N_{Training}$	44	-
	R^2	0.74	>0.5
Internal	Q^2_{LOO}	0.65	>0.5
	$r^2_{m(LOO)Scaled}$	0.54	>0.5
	$\Delta r^2_{m(LOO)Scaled}$	0.13	<0.2
	N_{Test}	15	-
	Q^2_{F1} *or* R^2_{pred}	0.73	>0.5
External	Q^2_{F2}	0.73	>0.5
	$r^2_{m(test)Scaled}$	0.64	>0.5
	$\Delta r^2_{m(test)Scaled}$	0.12	<0.2
	r^2	0.73	>0.5
	$\left| r_0^2 - r_0'^2 \right|$	0.05	<0.3
Golbraikh and Tropsha's criteria	$\frac{r^2 - r_0^2}{r^2}$	0.002	Any of them must be < 0.1
	$\frac{r^2 - r_0'^2}{r^2}$	0.06	
	k	1.01	$0.85 \leq k \, or \, k' \leq 1.15$
	k'	0.91	

Considering the complexity and diversity of FDs structures, the internal and external prediction variances are 0.74 and 0.73, respectively which are highly acceptable in QSPR modeling. The same value for Q^2_{F1} and Q^2_{F2} suggested stability between training and test sets followed by identical distribution of FDs. The model also passed the strict r_m^2 metrics and Golbraikh and Tropsha's criteria. To check the randomness of the model which one support that the model is not developed by chance, we have performed process validation by generation of 100 random models. We found that average R^2 and Q^2 values for 100 random models are 0.17 and -0.38, respectively which failed the stipulated value of 0.5 for both metrics. It supports the conclusion that the PLS model is not a result of correlation-by-chance. To check the applicability domain (AD), we have prepared ED-based AD study and found that all test compounds are falling within the AD zone created by the training set data which supports the reliability of prediction for test compounds.

3.2. Mechanistic Interpretation of Model

Out of seven features, two features namely S_A(chg)/A_D_D_D/1_2s,1_3s,3_4a/6 and S_A(type)/C.3_C.3_C.3_C.AR/1_3s,2_3s,3_4s/5 contributed positively to PCE value. This signifies that the higher value of these features may increase the PCE value. On the contrary, the remaining five features affect the equation negatively suggesting lowering or no effect on the PCE value (absent of these features or fragments in the FDs).

S_A(chg)/A_D_D_D/1_2s,1_3s,3_4a/6 defines a four-atomic fragment labeled by partial charges which are induced by -ortho directing groups in the benzene rings. FDs having the mentioned fragments (see Figure 1) have higher value for this descriptor and in a consequence promote higher PCE value. While, S_A(type)/C.3_C.3_C.3_C.AR/1_3s,2_3s,3_4s/5 represents types of fullerene substituent connections demonstrated in Figure 2 are also good for increment of PCE value. This specific fragment portrayed that aromatic rings like phenyl, thiophene, pyrrole (electron acceptors) attached to the fullerene by a linker help the electron withdrawing capability of the fullerene. Fr5(chg)/B_C_C_C_D/1_4s,2_3s,2_4s,3_4s/ also defines partial charges portrayed by the molecular fragment in Figure 3 offers detrimental effects to PCE. This suggests that it is better to avoid such specific substituents to FDs. Fr5(type)/C.3_C.3_C.3_C.3_H/1_2s,2_3s,3_4s,4_5s/ describes the existence of saturated carbon chains like [C(sp^3)- C(sp^3)- C(sp^3)- C(sp^3)-H]) and offers inductive effects and reduces the mesomeric effects of aromatic rings in a FD, which has a negative effect on PCE. Fr5(att)/C_C_E_E_E/1_3s,2_4s,3_5a,4_5a/ is associated to the van der Waals attraction between 3 or higher-ortho substituents in benzene rings with negative contribution to the PCE. Fr5(type)/C.3_C.3_C.AR_C.AR_C.AR/1_4s,2_3s,2_5s,4_5a/ portrayed substituents to a pentagon of the fullerene core which is electrochemically more steady than general two-points substituted FDs. Higher number of attachments in the parent fullerene affects the unsaturation and aromaticity negatively and results in reduction of acceptor property of FDs. ASA_P defines the solvent accessible surface area of polar atoms which is significant for the calculation of free energy changes as a result of shifting the molecule from a polar to a non-polar solvent during the formation of PSCs with BHJ layers.

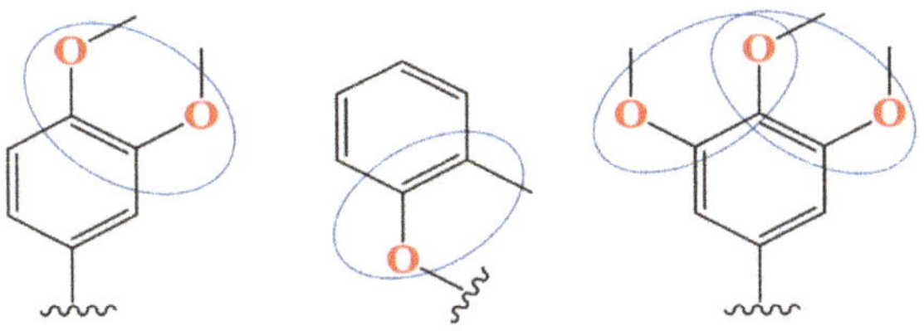

Figure 1. Fragments like -ortho directing groups substituted in the benzene ring help in power conversion efficiency (PCE) increment.

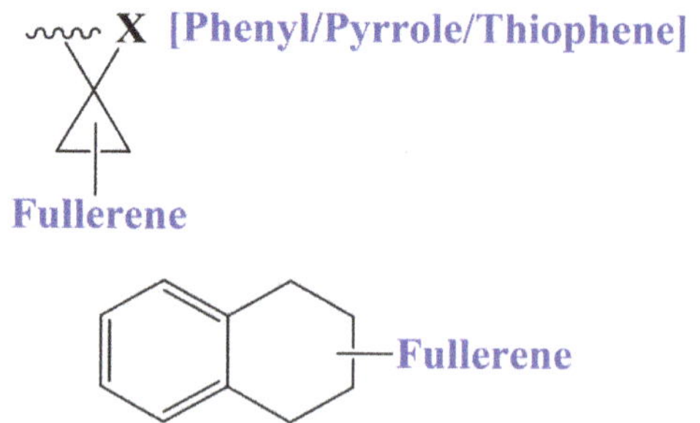

Figure 2. Important fullerene substituents for higher PCE.

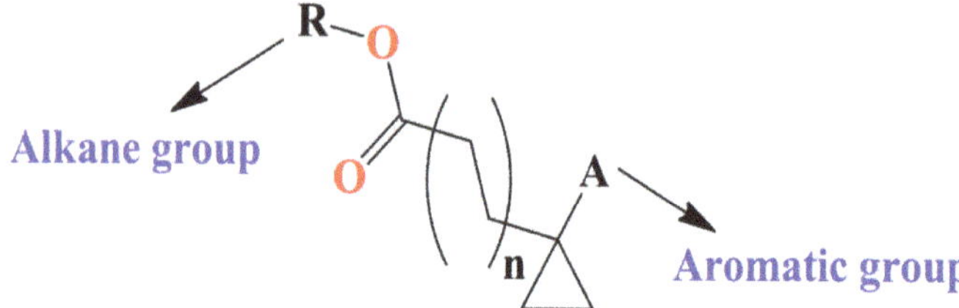

Figure 3. Fragment has detrimental effect on PCE value.

3.3. Designing of Novel FDs as Acceptor

Mechanistic interpretation of important fragments lead to higher PCE are considered here for the designing of lead FDs as acceptor for Fullerene-based PSCs. The major fragments illustrated in Figures 1 and 2 are included in the acceptor and fragments (Figure 3, $C(sp^3)$- $C(sp^3)$- $C(sp^3)$- $C(sp^3)$-H, 3 or higher-ortho substituents in benzene and substituents to a pentagon of the fullerene core) which are detrimental to PCE are avoided when possible. Ten FDs structures (7 C_{60} and 3 C_{70}) are designed (Figure 4) and modeled descriptors are calculated following similar protocols mentioned in Section 2.1.2. Followed by QSPR model (Equation (1)) is implemented to predict the PCE of the designed FDs and AD study has also been performed to check their prediction reliability. All 10 FDs passed the Euclidean distance-based AD study portraying the PCE values to be reliable and can be considered for the further introspection to prove them as future efficient acceptor for fullerene-based PSCs. The predicted PCE of FDs range from 7.96 to 23.01 considering both C_{60} and C_{70} FDs. Here, FD7, FD8 and FD9 are C_{70} FDs whose PCE values varies from 7.96 to 12.11, while remaining FDs are C60 FDs whose PCE values varies from 12.03 to 23.01. All values are no doubt encouraging and higher than any existing FD acceptor for PSCs. To check the electrochemical properties of these FDs, we have selected the top three C_{60} and top C_{70} FDs for further analysis. Computed modeled descriptors value along with mean normalized distance for AD study provided in Table S2 in the Supplementary material section.

3.4. Energetics of Isolated FDs

The computed HOMO energy of PCBM and P3HT are similar to experimental value while the LUMO values are overestimated by the chosen method which is accepted by the community [35]. According to McCormik et al. [35], B3LYP functional is sufficient to approximate HOMO energy of conjugated polymers while LUMO is not well approximated. Due to the overestimation of LUMO energy, the computed energy for HOMO-LUMO gap (E_{gap}) is very high with B3LYP functional. To tradeoff between LUMO energy and E_{gap}, we choose PBE functional for the ground state calculation. The computed value of the HL_g for the gas phase isolated PCBM is 2.82 eV but González et al. reported 1.48 eV (PBE-D3/TZP) [31] and by Thompson et al. it amounts to 1.4 eV [36] whereas Cook et al. reported experimental value for pure PCBM films to be 1.8 eV [34]. The computed energy gap for the isolated P3HT-8mer is 1.54 eV which is in good agreement with the reported results 1.31 eV [31] and 1.49 eV [37], whereas the experimental value for pure P3HT films is 1.9 eV [34]. Computed energetics of P3HT, PCBM and four FDs are compiled in Figure 5.

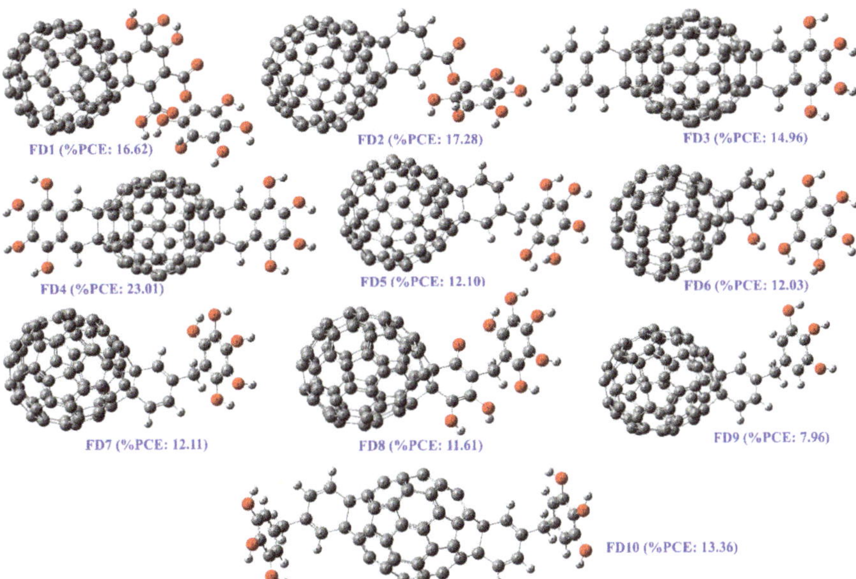

Figure 4. Structure of designed fullerene-derivatives (FDs) as lead acceptor molecule for polymer solar cells (PSCs).

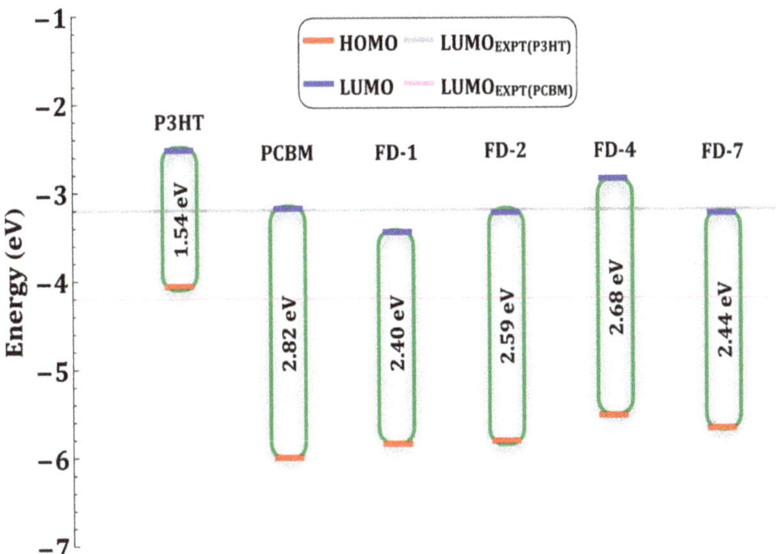

Figure 5. Computed energy diagram of the four FDs along with [6,6]-phenyl-C61-butyricacidmethylester (PCBM) and P3HT. All the values obtained with the use of PBE/6-31G(d,p) level of theory in the gas phase.

We also compute the different driving forces for the exciton binding, dissociation, charge-transfer, and open-circuit voltage (V_{OC}) for FDs which affect the smooth flow of exciton in the donor/acceptor blends. The following definition used to compute driving forces in terms of energy [38]:

$$\Delta E_1 = E_{Donor}^{LUMO} - E_{Acceptor}^{LUMO} \tag{2}$$

$$\Delta E_2 = E_{Acceptor}^{LUMO} - E_{Donor}^{HOMO} = V_{OC} \tag{3}$$

$$\Delta E_3 = E_{Donor}^{HOMO} - E_{Acceptor}^{HOMO} \tag{4}$$

The difference between the LUMO levels of donor and acceptor, ΔE_1, is responsible for the charge dissociation of the excitons in polymer donor to overcome the excitations binding energy. The typical exciton binding energy is ca. 0.3–0.5 eV, if it is too large then the exciton charge separation will require more energy and lowers V_{OC}. The value of ΔE_2 determines the V_{OC} which can be increased by up-shifted LUMO energy level of acceptors and thus higher efficiency of PSCs. ΔE_3, the difference between the HOMO levels of donor and acceptor, affects the dissociation of electron-hole pairs in the donor/acceptor interface. If the HOMO levels of acceptor are too high, ΔE_3 will be too small which hinders the dissociation of electron-hole pairs at interface in some extent [38]. So, it is necessary to maintain effective ΔE_3 to maintain a smooth dissociation. But ΔE_3 is not the sole factor which affects the efficiency of PSCs. Balancing between all the related factors (Table 3), FD4 will be efficient acceptor as it has the lower exciton binding energy and higher V_{OC} with least ΔE_3 value.

Table 3. Electronic energy level differences of P3HT and fullerene-derivatives (FDs) including [6,6]-phenyl-C61-butyricacidmethylester (PCBM).

FDs	ΔE_1	ΔE_2	ΔE_3
PCBM	0.66	0.88	1.94
FD1	0.93	0.61	1.79
FD2	0.71	0.83	1.76
FD4	0.33	1.21	1.47
FD7	0.72	0.82	1.62

3.5. UV-Vis Absorption Spectra of Isolated FDs

The Figure 6 shows the simulated absorption spectrum of the different PCBM, four designed FDs and P3HT. The spectrum of PCBM reproduces the qualitatively main features of reported experimental results [34,39], such as, it shows a strong optical absorption predominantly in the UV region, with very weak absorption (f = 0.0021) in the visible region (from 450 nm to 700 nm, see the inset in Figure 6). However, the spectrum of P3HT oligomer (Figure 6, Bottom) showing two strong absorption peaks at 241.20 nm (f = 0.3423) and 381 nm (f = 2.4886) with one shoulder at 307 nm (f = 0.4027), which represents a HOMO − LUMO + 6, HOMO − LUMO and HOMO − 1 − LUMO + 1 transition, respectively. As a reference, for pure P3HT films two maxima absorption peaks and one shoulder at 493 nm, 517 nm, and 572 nm, respectively have experimentally been reported, also attributed to the π-π* transitions [40–42]. Our simulated absorption of P3HT oligomer shows a small blue shift compare to experimental one.

Also, PCBM has weak absorption in visible region which is one of the factors that can be tuned to get better efficiency from PSCs [43]. If one examines the absorption strength of FDs, it is obvious that except FD4 and FD7 all of them having very weak absorption in the visible region. C_{70} derivatives FD7 is showing a red shift extends up to 600 nm with large oscillator strength [43]. From this aspect our designed FD4 and FD7 will be the most efficient acceptor in conjunction with donor P3HT.

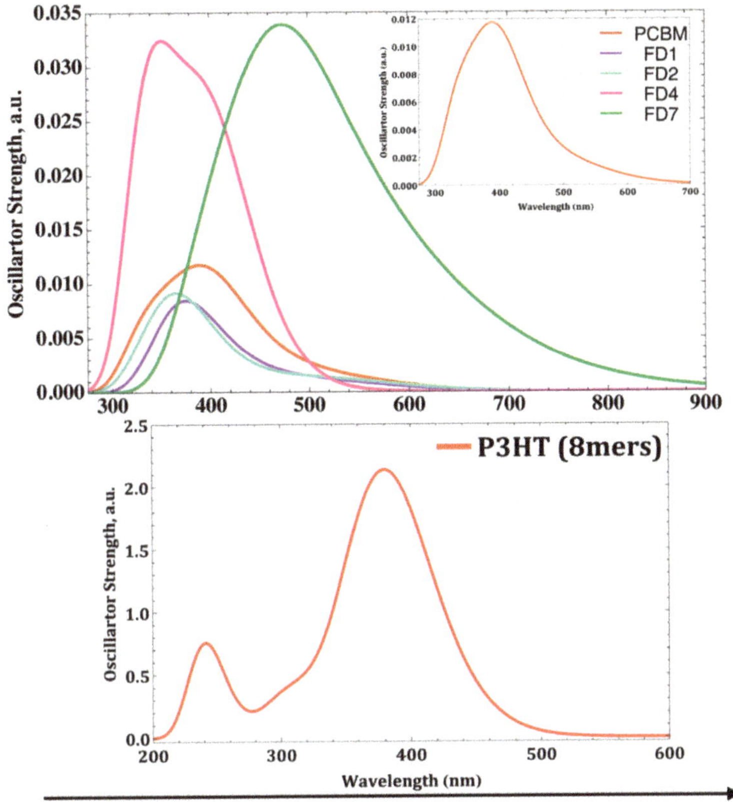

Figure 6. Simulated absorption spectra of PCBM, four FDs [Top] and P3HT [Bottom] with the use of TD/CAM-B3LYP/6-311G(d,p) level of theory in chlorobenzene solvent. Inset Top: Magnified PCBM absorption spectrum.

4. Conclusions

In silico modeling followed by designing and optoelectronic properties evaluation of future lead FDs as acceptor for PSCs had been performed. The QSPR model led us to development of 10 novel FDs as acceptor including seven C_{60} and three C_{70}. Based on the predicted PCE, optoelectronic properties of four FDs were evaluated by DFT and TDDFT. PBE/6-31G(d,p) and CAM-B3LYP/6-311G(d,p) level of theory were employed for gas phase DFT and solvent phase TDDFT computations. Frontier orbital energies and UV-Vis absorption spectra of the isolated P3HT oligomer, PCBM and FDs were analyzed to estimate the optoelectronic properties of four FDs as acceptor in future PSCs. Exciton binding energy plays the pivot role at interface when excitons diffuse and dissociate in to electrons on LUMO level of the acceptor. The big off-set of LUMO energy levels will hinders this process. FD4 is the best C_{60}-derivatives candidates for PSCs as it has the lowest exciton binding energy, up-shifted LUMO energy level that assist to increase V_{OC} and strong absorption in the UV region. In case of C_{70}-derivatives, FD7 is a potential candidate for future PSCs due to its strong absorption in UV-Vis region and lower exciton binding energy with higher V_{OC}. By trading off the computed optoelectronic properties, our analysis supports our QSPR model which predict highest PCE values for FD4.

The rational molecular modeling, designing, and prediction followed by quantum study offers valued reasoning for the synthesis of lead FDs with higher power conversion efficiency. The structural analysis concluded the following points:

- Ortho directing groups in the benzene rings and aromatic rings like phenyl, thiophene, pyrrole attached to the fullerene are significant features for better PCE of PSCs.
- Saturated carbon chains, 3 or higher *-ortho* substituents in benzene rings and a higher number of attachments in the parent fullerene core need to be avoided for higher PCE along with structural fragments with a lower solvent accessible surface area of polar atoms.

Supplementary Materials: The following are available online at http://www.mdpi.com/1996-1944/12/14/2282/s1, Table S1: Fullerene derivatives with their experimental and predicted % PCE, Table S2: Computed value of modeled descriptors for each FDs along with their predicted % PCE and mean normalized distance value.

Author Contributions: S.K., J.K.R., and J.L. conceived and designed the experiments; S.K. and J.K.R. performed the experiments; S.K., J.K.R., and J.L. analyzed the data; S.K. and J.K.R. contributed analysis tools; S.K., J.K.R., and J.L. wrote the paper.

Funding: Authors thankful to the Department of Energy (grant number: DE-SC0018322) and the NSF EPSCoR (grant number: OIA-1757220) for financial support.

Acknowledgments: Authors want to acknowledge the Extreme Science and Engineering Discovery Environment (XSEDE) by National Science Foundation grant number OCI-1053575 and XSEDE award allocation number DMR110088 and DMR110013P for providing state-of-the-art high-performance computing facilities for supporting this research.

Conflicts of Interest: The authors declare no conflict of interest.

References

1. Li, S.; Ye, L.; Zhao, W.; Yan, H.; Yang, B.; Liu, D.; Li, W.; Ade, H.; Hou, J. A wide band gap polymer with a deep highest occupied molecular orbital level enables 14.2% efficiency in polymer solar cells. *J. Am. Chem. Soc.* **2018**, *140*, 7159–7167. [CrossRef] [PubMed]

2. Cui, Y.; Yao, H.; Zhang, J.; Zhang, T.; Wang, Y.; Hong, L.; Xian, K.; Xu, B.; Zhang, S.; Peng, J.; et al. Over 16% efficiency organic photovoltaic cells enabled by a chlorinated acceptor with increased open-circuit voltages. *Nat. Commun.* **2019**, *10*, 2515. [CrossRef] [PubMed]

3. Chang, Y.; Wu, C.E.; Chen, S.Y.; Cui, C.; Cheng, Y.J.; Hsu, C.S.; Wang, Y.-L.; Li, Y. Enhanced performance and stability of a polymer solar cell by incorporation of vertically aligned, Cross–Linked fullerene nanorods. *Angew. Chem. Int. Ed.* **2011**, *50*, 9386–9390. [CrossRef] [PubMed]

4. Meng, L.; Zhang, Y.; Wan, X.; Li, C.; Zhang, X.; Wang, Y.; Ke, X.; Xiao, Z.; Ding, L.; Xia, R.; et al. Organic and solution-processed tandem solar cells with 17.3% efficiency. *Science* **2018**, *361*, 1094–1098. [CrossRef] [PubMed]

5. Vanlaeke, P.; Vanhoyland, G.; Aernouts, T.; Cheyns, D.; Deibel, C.; Manc, J.; Heremansa, P.; Poortmans, J. Polythiophene based bulk–heterojunction solar cells: Morphology and its implications. *Thin Solid Films* **2006**, *511–512*, 358–361. [CrossRef]

6. Chi, D.; Qu, S.; Wang, Z.; Wang, J. High efficiency P3HT: PCBM solar cells with an inserted PCBM layer. *J. Mater. Chem. C* **2014**, *2*, 4383–4387. [CrossRef]

7. Tsai, J.H.; Lai, Y.C.; Higashihara, T.; Lin, C.J.; Ueda, M.; Chen, W.C. Enhancement of P3HT/PCBM photovoltaic efficiency using the surfactant of triblock copolymer containing Poly(3-hexylthiophene) and Poly(4-vinyltriphenylamine) segments. *Macromolecules* **2010**, *43*, 6085–6091. [CrossRef]

8. Zhou, H.; Zhang, Y.; Seifter, J.; Collins, S.D.; Luo, C.; Bazan, G.C.; Nguyen, T.-Q.; Heeger, A.J. High-efficiency polymer solar cells enhanced by solvent treatment. *Adv. Mater.* **2013**, *25*, 1646–1652. [CrossRef] [PubMed]

9. Winder, C.; Sariciftci, N.S. Low bandgap polymers for photon harvesting in bulk heterojunction solar cells. *J. Mater. Chem.* **2004**, *14*, 1077–1086. [CrossRef]

10. Mühlbacher, D.; Scharber, M.; Morana, M.; Zhu, Z.; Waller, D.; Gaudiana, R.; Brabec, C. High photovoltaic performance of a low-bandgap polymer. *Adv. Mater.* **2006**, *18*, 2884–2889. [CrossRef]

11. Zhang, M.; Gu, Y.; Guo, X.; Liu, F.; Zhang, S.; Huo, L.; Russell, T.P.; Hou, J. Efficient polymer solar cells based on benzothiadiazole and alkylphenyl substituted benzodithiophene with a power conversion efficiency over 8%. *Adv. Mater.* **2013**, *25*, 4944–4949. [CrossRef] [PubMed]

12. Park, S.H.; Roy, A.; Beaupré, S.; Cho, S.; Coates, N.; Moon, J.S.; Moses, D.; Leclerc, M.; Lee, K.; Heeger, A.J. Bulk heterojunction solar cells with internal quantum efficiency approaching 100%. *Nat. Photonics* **2009**, *3*, 297–302. [CrossRef]

13. Albrecht, S.; Janietz, S.; Schindler, W.; Frisch, J.; Kurpiers, J.; Kniepert, J.; Inal, S.; Pingel, P.; Fostiropoulos, K.; Koch, N.; et al. Fluorinated copolymer PCPDTBT with enhanced open-circuit voltage and reduced recombination for highly efficient polymer solar cells. *J. Am. Chem. Soc.* **2012**, *134*, 14932–14944. [CrossRef] [PubMed]

14. Roy, K.; Kar, S.; Das, R.N. *Understanding the Basics of QSAR for Applications in Pharmaceutical Sciences and Risk Assessment*; Academic Press: Cambridge, MA, USA, 2015.

15. Kar, S.; Sizochenko, N.; Ahmed, L.; Batista, V.S.; Leszczynski, J. Quantitative structure-property relationship model leading to virtual screening of fullerene derivatives: Exploring structural attributes critical for photoconversion efficiency of polymer solar cell acceptors. *Nano Energy* **2016**, *26*, 677–691. [CrossRef]

16. Kar, S.; Roy, J.; Leszczynska, D.; Leszczynski, J. Power conversion efficiency of Arylamine organic dyes for dye-sensitized solar cells (DSSCs) explicit to cobalt electrolyte: Understanding the structural attributes using a direct QSPR approach. *Computation* **2017**, *5*, 2. [CrossRef]

17. Kar, S.; Roy, J.K.; Leszczynski, J. In silico designing of power conversion efficient organic lead dyes for solar cells using todays innovative approaches to assure renewable energy for future. *NPJ Comput. Mater.* **2017**, *3*, 22. [CrossRef]

18. Li, H.; Zhong, Z.; Li, L.; Gao, R.; Cui, J.; Gao, T.; Hu, L.H.; Lu, Y.; Su, Z.M.; Li, H. A cascaded QSAR model for efficient prediction of overall power conversion efficiency of all-organic dye-sensitized solar cells. *J. Comput. Chem.* **2015**, *36*, 1036–1046. [CrossRef]

19. Venkatraman, V.; Alsberg, B.K. A quantitative structure-property relationship study of the photovoltaic performance of phenothiazine dyes. *Dyes Pigment.* **2015**, *114*, 69–77. [CrossRef]

20. Dennington, R.; Keith, T.A.; Millam, J.M. *GaussView, Version 6*; Semichem Inc.: Shawnee Mission, KS, USA, 2016.

21. Frisch, M.J.; Trucks, G.W.; Schlegel, H.B.; Scuseria, G.E.; Robb, M.A.; Cheeseman, J.R.; Scalmani, G.; Barone, V.; Petersson, G.A.; Nakatsuji, H.; et al. *Gaussian 16, Revision, B.01*; Gaussian, Inc.: Wallingford, CT, USA, 2016.

22. Chemaxon Software. Available online: https://www.chemaxon.com (accessed on 1 May 2019).

23. QSAR4U Software. Available online: http://qsar4u.com/pages/sirms.php (accessed on 1 May 2019).

24. Artemenko, A.G.; Muratov, E.N.; Kuz'min, V.E.; Muratov, N.N.; Varlamova, E.V.; Kuz'mina, A.V.; Gorb, L.G.; Golius, A.; Hill, F.C.; Leszczynski, J. QSAR analysis of the toxicity of nitroaromatics in Tetrahymena pyriformis: Structural factors and possible modes of action. *SAR QSAR Environ. Res.* **2011**, *22*, 575–601. [CrossRef]

25. DTC Lab Software Tools. Available online: http://teqip.jdvu.ac.in/QSAR_Tools/ (accessed on 1 May 2019).

26. Roy, K.; Mitra, I.; Kar, S.; Ojha, P.; Das, R.N.; Kabir, H. Comparative studies on some metrics for external validation of QSPR models. *J. Chem. Inf. Model.* **2012**, *52*, 396–408. [CrossRef]

27. Schüürmann, G.; Ebert, R.-U.; Chen, J.; Wang, B.; Kühne, R. External validation and prediction employing the predictive squared correlation coefficient test set activity mean vs. training set activity mean. *J. Chem. Inf. Model.* **2008**, *48*, 2140–2145. [CrossRef] [PubMed]

28. Golbraikh, A.; Tropsha, A. Beware of q^2! *J. Mol. Graph. Model.* **2002**, *20*, 269–276. [CrossRef]

29. Kar, S.; Roy, K.; Leszczynski, J. Applicability Domain: A Step towards Confident Predictions and Decidability for QSAR Modeling. In *Computational Toxicology: Methods and Protocols, Methods in Molecular Biology*; Nicolotti, O., Ed.; Springer Nature: Basingstoke, UK, 2018; Volume 1800, pp. 395–443.

30. Roy, K.; Kar, S. How to Judge Predictive Quality of Classification and Regression Based QSAR Models? In *Frontiers of Computational Chemistry*; Haq, Z.U., Madura, J., Eds.; Bentham Science Publishers: Sharjah, UAE, 2015; pp. 71–120.

31. Gutiérrez-González, I.; Molina-Brito, B.; Götz, A.W.; Castillo-Alvarado, F.L.; Rodríguez, J.I. Structural and electronic properties of the P3HT–PCBM dimer: A theoretical Study. *Chem. Phys. Lett.* **2014**, *612*, 234–239. [CrossRef]

32. Martínez, J.P.; Trujillo-González, D.E.; Götz, A.W.; Castillo-Alvarado, F.L.; Rodríguez, J.I. Effects of Dispersion Forces on Structure and Photoinduced Charge Separation in Organic Photovoltaics. *J. Phys. Chem. C* **2017**, *121*, 20134–20140. [CrossRef]

33. Cossi, M.; Barone, V. Time-dependent density functional theory for molecules in liquid solutions. *J. Chem. Phys.* **2001**, *115*, 4708. [CrossRef]

34. Cook, S.; Katoh, R.; Furube, A. Ultrafast Studies of Charge Generation in PCBM: P3HT Blend Films following Excitation of the Fullerene PCBM. *J. Phys. Chem. C* **2009**, *113*, 2547–2552. [CrossRef]

35. McCormick, T.M.; Bridges, C.R.; Carrera, E.I.; DiCarmine, P.M.; Gibson, G.L.; Hollinger, J.; Kozycz, L.M.; Seferos, D.S. Conjugated Polymers: Evaluating DFT Methods for More Accurate Orbital Energy Modeling. *Macromolecules* **2013**, *46*, 3879–3886. [CrossRef]

36. Thompson, B.C.; Fréchet, J.M.J. Polymer-fullerene composite solar cells. *Angew. Chem. Int. Ed. Engl.* **2008**, *47*, 58–77. [CrossRef]

37. Drori, T.; Sheng, C.-X.; Ndobe, A.; Singh, S.; Holt, J.; Vardeny, Z.V. Below-Gap Excitation of π-Conjugated Polymer-Fullerene Blends: Implications for Bulk Organic Heterojunction Solar Cells. *Phys. Rev. Lett.* **2008**, *101*, 37401. [CrossRef]

38. He, Y.; Li, Y. Fullerene derivative acceptors for high performance polymer solar cells. *Phys. Chem. Chem. Phys.* **2011**, *13*, 1970–1983. [CrossRef]

39. Cook, S.; Ohkita, H.; Kim, Y.; Benson-Smith, J.J.; Bradley, D.D.C.; Durrant, J.R. A photophysical study of PCBM thin films. *Chem. Phys. Lett.* **2007**, *445*, 276–280. [CrossRef]

40. Li, G.; Shrotriya, V.; Yao, Y.; Huang, J.; Yang, Y. Manipulating regioregular poly(3-hexylthiophene): [6,6]-phenyl-C61-butyric acid methyl ester blends—route towards high efficiency polymer solar cells. *J. Mater. Chem.* **2007**, *17*, 3126. [CrossRef]

41. Shrotriya, V.; Ouyang, J.; Tseng, R.J.; Li, G.; Yang, Y. Absorption spectra modification in poly(3-hexylthiophene): Methanofullerene blend thin films. *Chem. Phys. Lett.* **2005**, *411*, 138–143. [CrossRef]

42. Yin, B.; Yang, L.; Liu, Y.; Chen, Y.; Qi, Q.; Zhang, F.; Yin, S. Solution-processed bulk heterojunction organic solar cells based on an oligothiophene derivative. *Appl. Phys. Lett.* **2010**, *97*, 023303. [CrossRef]

43. He, Y.; Zhao, G.; Peng, B.; Li, Y. High-yield synthesis and electrochemical and photovoltaic properties of indene-C 70 bisadduct. *Adv. Funct. Mater.* **2010**, *20*, 3383–3389. [CrossRef]

MDPI
St. Alban-Anlage 66
4052 Basel
Switzerland
Tel. +41 61 683 77 34
Fax +41 61 302 89 18
www.mdpi.com

Materials Editorial Office
E-mail: materials@mdpi.com
www.mdpi.com/journal/materials